The Science of
The Hitchhiker's Guide to the Galaxy

The Science of
The Hitchhiker's Guide to the Galaxy

Michael Hanlon

Macmillan

First published 2005 by
Macmillan
Houndmills, Basingstoke, Hampshire RG21 6XS and
175 Fifth Avenue, New York, N. Y. 10010
Companies and representatives throughout the world

ISBN-13: 978–1–4039–4577–8 hardback
ISBN-10: 1–4039–4577–2 hardback
ISBN-13: 978–1–4039–9726–5 paperback
ISBN-10: 1–4039–9726–8 paperback

This book is printed on paper suitable for recycling and made from fully managed
and sustained forest sources.

A catalogue record for this book is available from the British Library.

A catalog record for this book is available from the Library of Congress.

10   9   8   7   6   5   4   3   2   1
14   13   12   11   10   09   08   07   06   05

Printed and bound in the United States of America

contents

acknowledgements

First of all thanks must go to Sara Abdulla at Macmillan for coming up with such a brilliant idea, for thinking of me and for persuading me to do it. Given the nature of this beast I have been at the mercy of the People Who Know, and, as ever, it seems that there are an infinite number of scientists out there willing to lend an author a hand in his darkest hour. Particular thanks must go to Fred Adams who put me right over the fate of the Universe, and to Martin Rees who set me straight about the multiverse. I must also give thanks to David Harland for his valuable and face-saving comments on the book, to my agent Simon Trewin at PFD and, of course, to Elena Seymenliyska for her continuing love and support. Finally, I am indebted to the late Douglas Adams for, well, everything. I never met him, but he changed the way I thought about life, and for that I will always be profoundly grateful.

MH February 2005

1

introduction

Spawned from the radio series of the same name, *The Hitch-hiker's Guide to the Galaxy* novels are a whimsical, sometimes cutting, satire on science fiction, life, the Universe and indeed everything. Created by Douglas Adams, a former scriptwriter for the television series *Dr Who*, the 'increasingly inaccurately named' trilogy eventually ran to five volumes. It charts the bizarre escapades of Arthur Dent, a hapless BBC employee rescued by his friend Ford Prefect just as aliens called Vogons demolish Earth to make way for a hyperspace bypass. What follows is a colourful and frequently ludicrous saga in which Dent, Earth's Everyman, is exposed to everything the galaxy can throw at him. Initially played for a knowing laugh, the series becomes darker, ending with the positively bleak *Mostly Harmless*, in which closure, of sorts, is achieved.

The Hitchhiker's Guide is, on one level, a comedy. Much of the 'science' therein is clearly bonkers, and intentionally so. One-time Galactic president Zaphod Beeblebrox has (usually) two heads and three arms. There are improbable planets where it is permanently Saturday afternoon; flying office blocks; a mostly ocean world with an island called, for some reason, France; and a spacecraft marooned for aeons where paralysed passengers are woken every century to be served refreshments while the crew await delivery of lemon-soaked paper napkins. There are triple-breasted whores and an articulate bovine bred to want to be eaten. Yet something else nestles among the gentle mockery of *Who*-style japes, space operas like *Star Wars* and more serious stuff from the pens of Isaac Asimov, Arthur C. Clarke and the rest. The *Guide* records Douglas Adams's fascination with the increasingly strange twists and turns of cutting-edge cosmology and theoretical physics.

The first parts of the *Guide* were written in 1978 when black holes, parallel universes, quantum weirdness and serious debates on alien life first took a real hold on the public consciousness. So despite the whimsy, *Hitchhiker's* bristles with real science and technology. Adams caught the wave of interest in the New Cosmology – Big Bangs, black holes and Grand Unified Theories – and surfed it with wit. Had he never posed the Ultimate Question of Life, the Universe and Everything, would Stephen Hawking have sold all those books attempting to answer it?

Some of the most interesting themes of the series concern parallel universes and alternate realities. The idea that another, shadow, world lies just round the corner is haunting. Humans, especially young ones, have always been drawn to the notion of an elsewhere – at the bottom of the garden, the back of the wardrobe, the top of the Faraway Tree or from Platform 9¾. Then came the shock that fantastical elsewheres may be the only explanation for the wonderland that is the quantum world.

It is this world that holds the key to some of the weirdest nooks and crannies in the very strange universe that science is uncovering. The land of the quantum is that of the most minute and brief. This is where electrons are not distinct entities but fuzzy wave functions which can be the size of the entire universe or so small that a trillion can be crammed on the point of a needle. Things can be in two places at once, and simply looking at something affects it in a profound and deeply strange way.

The quantum world could, reckon some researchers, open the door to true parallel universes: worlds where D-Day failed and Hitler won the Second World War, worlds where there never *was* a Second World War and where the final Ford Sierra off the production line was painted just a *slightly* less fetching shade of metallic violet. That is the great thing about parallel universes: they can be as different from or similar to ours as you please. Poor Arthur Dent finds and loses the love of his life thanks to an unfortunate tryst with the push–pull of parallels. Trillian, queen of the galactic airwaves and possibly the Universe's least competent mother, confronts her parallel self – a hard-working TV journalist whose feet had only once left the Earth.

Like all the best science fiction, the *Guide* is as much about philosophy as it is about science. Are our lives predetermined? Do we live in an essentially billiard-ball, Newtonian universe where – despite the awkward fuzziness of chaos theory and quantum mechanics – each twist and turn of our future is predictable to infinite precision? Or is the future unknowable both in practice and in theory? Told he will experience his *dénouement* in a place he's never heard of called Stavro Mueller Beta at an unspecified time and date, Arthur is naturally alarmed. We like to think that the past is set in stone, the future just a fog.

To cope with this uncertainty and the alarming reality of our mortality, some – including the super-intelligent mice of the *Guide* – have decided they are striving for the Ultimate Answer to the Ultimate Question. But as pointed out by Deep Thought (the mega-computer built by the mice to find out what is *really* going on), the Answer, whatever it is, will have no meaning until we work out what that Question is.

Which brings us neatly to God. The *Guide* was begun just before reason trotted to the back of the bus and had a quiet smoke, leaving the front seats to be occupied by the gibbering prophets of the New Age. When Adams wrote his original scripts, traditional Church-based faith was on the wane. Then along came the crystals and the tarot cards, the chakras and the wicca women.

This was a rerun of the old Victorian spiritualism, another, essentially decadent, *fin de siècle* fad. New Age belief, with its emphasis on Celtic or Oriental mysticism, self awareness and non-Western approaches to mental and physical health, diet and medicine, was a sort of religion-lite, a reaction to 1960s materialism and technological triumphalism. To slightly misquote G. K. Chesterton: when we stopped believing in God, we didn't start believing in nothing, but came instead to believe in everything.

According to the *Guide*, the Babel fish, possibly the most extraordinary creature ever to have (allegedly) evolved, is proof enough of the non-existence of God. This small creature sits in your ear and provides a perfect simultaneous translation of any language in the Universe. It is so mind-bogglingly useful and improbable that it seems to prove the presence of intelligent design – thus refuting such proof in an instant, since God relies on faith, not logic, to keep Him going in the long dark teatimes of the soul. In *The Restaurant at the End of the Universe*, we meet the man who rules the cosmos. Except he is not God, just a rather confused individual who lives on a windswept beach with his cat, in a shed. He doesn't seem to know any more about what is really going on than anyone else.

The *Hitchhiker's* series articulates brilliantly the feeling that deep down, beneath it all, Things Are Not What They Seem. As Adams was writing, the idea was gaining ground that NASA had decided early on in the 1960s that a manned flight to the Moon was an impossibility given the paucity of technology at the time. Millions had begun to believe (some still do) that the lunar landings were faked and billions of dollars had been spent convincing the people of the world that America had won its race with Russia to put a human on the surface of another world.

Since 1978, certain scientific assumptions and fashions have been round the wheel of respectability and ridicule, sometimes more than once. Alien life, for instance, is taken for granted in the world of Zaphod, Ford and the Vogons. Yet in the 1970s, extraterrestrials, while in vogue in cinemas, were distinctly unfashionable in the lab. It was accepted wisdom that life was something very special. The early throes of the space age had turned up nothing, either on the dusty surface of the Moon or on Mars's marginally more inviting plains. The 19th-century vision – propounded by astronomers such as Percival Lowell – that the Universe was teeming with life had arguably prompted the entire space programme. This very programme, the astronauts and unmanned robots despatched from Baikonur and Canaveral, was turning up more and more evidence that we are alone. Scientists dubbed the arrival of *Mariner 4* at Mars in 1965 the 'Great Disappointment'. They knew that Lowell's hope of canals and oases was dead, but had hoped for some vestige of the lost Victorian dream. No princesses, perhaps, but maybe some lichens or shrubs?

It was not to be. Mars looked dead and Venus was more hellish than anyone had imagined. Adams's aliens reflect the astrobiological pessimism of the time. He makes little attempt to render them realistic. Most are humanoid, with funny foreheads or strange hairdos. They are the aliens of *Star Wars* and *Star Trek*, human ciphers onto which can be projected more extreme versions of human angst, violent emotions, bad poetry and politicking.

In the past twenty years or so the alien has undergone something of a revival. On Earth we have discovered that microbes can thrive in environments previously thought to be utterly untenable for life. These 'extremophiles' gobble sulphur for breakfast around deep-sea vents, and bask in temperatures which would sterilize a surgeon's instruments. And the latest results from the Mars probes have been more equivocal than those from the early *Mariners*. At the end of the 1970s the Red Planet was a cratered, dusty version of the Moon, with a near-vacuum atmosphere and sub-Antarctic climate. Now we see a world of ancient rivers and lakes, sedimentary layers and the promise of life – if not extant then at least extinct. Jupiter's moon Europa, a smooth blob of ice, probably has a vast ocean under its frozen surface. Inevitably, there is now speculation that microbes may lurk in this alien, inky abyss and – just possibly – beasts beyond imagining. Everywhere they look, it seems, scientists now see possible abodes of life.

The late cosmologist Carl Sagan cautioned against the 'chauvinism' which assumes that alien life must in the most essential ways resemble that here on Earth. *Star Trek* gave a conscious nod to this concept with its brilliant catch-phrase, 'It's life, Jim, but not as we know it'. Nonetheless, since the 1980s, the vastly improved computing resources of our civilization have been turned, in relatively minor but imaginative ways, to the search for any alien civilizations communicating much as we do. Microbes on Europa are all very well, but what about the little green men of sci-fi lore? SETI, the Search for Extraterrestrial Intelligence, uses spare time on the world's largest radio telescopes, and spare computing power on personal computers worldwide, to scan the heavens for signals broadcast by alien civilizations.

Occasionally, Douglas Adams managed not only to predict the future but also to create it. The Babel fish is now part of the world's cyberlexicon as the name of a website which performs translations – often hilariously badly. The *Guide*'s 'Sub-Etha Net' seems like an eerie premonition of the vast internet that runs our lives without us even noticing. 'Life, the Universe and Everything' is now the stock in trade of the punk-cosmologists and new philosophers. When Deep Thought announced that the answer to the said conundrum was '42', Adams was making a joke at philosophy's expense. Then, in 1999, Britain's Astronomer Royal, no less, wrote that the Universe does indeed boil down to six simple numbers.

Plenty of science and technology did not pan out as Douglas Adams mused. His alternate universe is essentially a place of technological optimism. Huge computers thrum silently in the background, acting simply as calculating engines of which Babbage would have been proud. Adams did not conceive of a place where machines with brains the size of a planet sit in every child's bedroom, bringing them into contact with pederasts and pornography. His cosmos is explored not by spacecraft with all the elegance of a flying washing machine or baroque, unsafe shuttles covered with china tiles, but by huge, grey hulking spaceships, sleek black cruisers and the wonderful, gorgeous *Heart of Gold*, the Lamborghini Miura of space travel, so fast that it can cross a galaxy in the wink of a coincidence. His future is, of course, that of 25 years ago.

And yet, and yet. So much that is in these wonderful books time and again bubbles up in the real future. Time itself, for instance. Back in the 1970s, time travel was thought to be essentially ridiculous. A few brave souls ventured that perhaps Einstein's equations did not actually preclude such a monstrosity, but physics generally coughed a polite cough and proclaimed that time travel should really be consigned to the world of fantasy, if not that of the plain silly. Then along came researchers such as Paul Davies, Kip Thorne and Stephen Hawking, who started asking awkward questions about the nature of space–time and whether or not it actually is possible to go back and shoot your own grandfather, or even to go back and *become* your own grandfather ....

Now physicists sit in their armchairs postulating what such a journey would require. Huge, spinning cylinders made of neutron matter, the stuff of pulsars, so dense that a teaspoonful would weigh as much as an aircraft carrier. Tame, whirling black holes parked a polite distance away from anyone who would be torn to bits or indeed offended by the obscene breakdown in causality that accompanies a naked singularity. Most bizarrely of all, physicists ponder how they could construct a wormhole, a tube of space–time linking anywhere with anywhen. The engineering challenges would be immense – there is talk of harnessing the power of a million galaxies, squashing objects the size of Jupiter into a suitcase, and setting off dozens of H-bombs in formation, like a gigantic and apocalyptic firework display. Actually, one man reckons he can pull off the time travel trick using just a couple of lasers in the lab. If he is right, time travel could be as easy as it is (or will be – with time travel, you have to watch your grammar as carefully as the equations of space–time) at the Restaurant at the End of the Universe.

Ah, Milliways! One of the most fantastical creations in the history of catering. A restaurant built on a ruined planet at which guests are entertained as the fires of cosmic Armageddon itself burn around them. The staff take your (fabulously expensive) orders at a time when time is about to run out. 'It's too late to worry about whether you left the gas on now', diners are told. The end of the Universe here is portrayed as a spectacular event: it ain't over till the last fat supernova sings.

Until quite recently, it seemed probable that time would indeed end with a bang – a rewind of the big one with which it all began. Gravity, that all-pervasive and most tricksy of forces, would eventually get the upper hand over the pesky expansion of the Universe. At some point in the far, but not infinitely far, future, the Big Bang would run out of steam. As if attached to elastic threads, the galaxies' outward expansion would slow down and eventually stop momentarily, followed by an ever-accelerating rushing together culminating in the Big Crunch. Oh, and at some point the sky would boil with the light of a billion suns.

Now the Big Crunch is off. Probably. Astronomers weigh the Universe with a fastidiousness that would put a dieters' convention to shame and they keep coming up with the wrong number. The place is simply too light to stop cosmic expansion. The Universe, they conclude, ends not with a bang but a whimper. Galaxies continue to rush apart, stars continue to be born and die, and eventually, after an extremely long but still finite time, the stuff starts to run out. In the ultimate fuel crisis, the raw matter powering the nuclear explosions that keep the lights burning becomes ever more scarce. Eventually, the lights go out for the very last time. The Universe is still full of stuff, but it is very, very black. Eventually, nothing interesting will ever happen again.

This news that cosmoskind (for we are all in this together, Betelgeusians, Vogons, Golgafrinchans and the rest) has ahead of it an eternity far bleaker than Hieronymous Bosch contemplated in his darkest of moods is grim. As grim, in fact, as the fate awaiting victims of the Total Perspective Vortex, given one momentary glimpse of the entire unimaginable infinity of the

Universe, and somewhere in it a tiny little marker, a nano-dot on a nano-dot, which says 'You are here'.

As the great biologist J. B. S. Haldane put it, 'The Universe isn't only queerer than we suppose, but queerer than we CAN suppose'. This suspicion lurks at the back of everyone's mind who has ever peered even a little under its bonnet. The Universe is big and supremely odd. It is odd that it is even here. It is odder still that it is here in a way that allows beings like us to have evolved to pose these impossible questions.

Today we know that the Universe is far stranger than even Douglas Adams supposed two decades ago. Most of it is very dangerous indeed; you really wouldn't want to be anywhere else out there that we know about rather than safely down here on Earth. Yet our bit of the cosmos is apparently perfectly tuned for life. To some, it has the smack of a set-up about it. To others, the myriad possibilities afforded by multiverse theory offer some sort of explanation.

Perhaps it *is* a set-up. Perhaps there is a god, but he is only joking. If so, maybe we are owed an apology.

2

where are the aliens?

Of all the races in the Galaxy who could have come and said a big hello to Earth, didn't it just have to be the Vogons.

Ford Prefect's thoughts minutes after the first unequivocal contact between Homo sapiens *and intelligent (albeit somewhat unsanitary) alien life.*

There are, it has to be said, an awful lot of places to live out there. Our medium-sized galaxy contains some 200 billion stars, and there are probably more galaxies than that in our observable Universe. In July 2003, scientists at a meeting of the International Astronomical Union in Australia announced their latest estimate of the number of stars in the Universe – 70 sextillion. That is 7 followed by a mind-boggling 22 zeros. It is bigger, for example, than the previous estimate, made by NASA, of just one sextillion stars (NASA's unofficial estimate of the number of stars in the Universe is 'zillions'). It is bigger even than the US budget deficit. It is bigger, possibly, than the price of an average house in London in whatever currency you choose to mention.

The new estimate means that the number of stars in the visible Universe is larger – quite a bit larger, actually – than the total number of all the grains of sand on all the beaches on Earth. And this estimate is almost certainly way too small. It takes in only the stars within range of our telescopes. It is quite possible that the Universe is vastly bigger than this horizon – which puts it at about 27 billion light years across. During the Universe's earliest moments space expanded, faster than light, into a vast volume that may be infinite. Most of the Universe could be so far away that we will never see it. All this is before we even start worrying about other universes or dimensions. All in all, the whole shebang is big.

Assuming, for a moment, that the kinds of places that life likes tend to be smallish rocky planets like our own (a big assumption, and one oft-criticized, but more of that later) orbiting medium-sized stable stars in well-behaved solar systems, we are still left with an awful lot of cosmic real estate. All the grains of sand in all the beaches in Florida, say. Or Crete.

Until the mid-1990s, the existence of planets orbiting other stars was widely assumed but unproven. Since then, more than 150 'extrasolar' planets have been found orbiting nearish stars, largely thanks to Earth's planet-finder-generals, the Californian astronomers Geoffrey Marcy and Paul Butler. One or two extrasolars were spotted because they, their parent stars and we are all in a line of sight, and we can see the faint lightening and darkening they cause as they partially eclipse their sun. The rest have been detected by an indirect means called Doppler spectroscopy. A large planet gives its parent star a gravitational nudge that makes it wobble slightly. This alters the wavelength of light the star appears to emit, which turns slightly bluer as the star wobbles towards us, and redder as it wobbles away. Since we are talking about objects several trillion kilometres away and wobbling distances of just a few hundred kilometres, that we can do this at all is testament to modern computers and spectrum analyzers.

Most extrasolar planets are huge – Jupiter-sized or above. Smaller planets are out there; they are just harder to find. As I write, a planet the size of Neptune has just been discovered. Many of the new-found solar systems are also pretty weird.

Most feature gas giants just a few million kilometres from their stars, much nearer to their parent stars than Mercury is to our own Sun. Again, this is a consequence of current detection techniques. As computers and telescopes improve, astronomers will undoubtedly discover more familiar types of solar system.

So we know that there are plenty of stars out there. And probably plenty of planets too. Trillions, quadrillions – hell – sextillions. Many will be too hot or too cold for life, but many – trillions, or at least a Malibu's-worth of sand grains – will be just right. Say, well, one per thousand solar systems. Oh all right, one per million. Whatever.

So, where *is* everybody?

The physicist Enrico Fermi put this paradox to a group of fellow scientists over lunch at Los Alamos in 1950. It was becoming clear that the Universe was vastly larger than anyone had suspected: astronomers discovered in the 1920s that *nebulae*, previously thought to be clouds of gas within our galaxy, were in fact separate galaxies. That same decade brought the realization that the Universe is billions of years old – ten times older than previously thought. By 1950, stars and planets, however distant, were known to be made of the same materials and to obey the same physical laws that pertain closer to home. No one had discovered any extrasolar planets by that stage, of course, but it was logical to assume they were there.

A Universe of such size and antiquity, the lunchers agreed, should be teeming with life, some of it intelligent like us. And some of this intelligent life must have created technological civilizations millions or billions of years old. It is inconceivable that none of these super-civilizations would have found the will or wherewithal to explore space. In our galaxy alone there must be thousands – perhaps tens of thousands or millions – of such star-farers. So, where are they?

The arrival of the Vogons in the first instalment of the *Guide* illustrates how many people believe this paradox will be resolved. Aliens are out there, the reasoning goes, a lot of them, and the only reason they haven't found us yet is that we've simply not been noticed. In time, we will be, and when we do the results will be dramatic. Here's hoping that the first aliens turn out to be nothing like the ghastly Vogons ....

A more alarming number of people believe that, despite space's deafening silence, aliens have *already* made their presence felt. The UFO theory is one well-established solution to the Fermi Paradox. Sadly it does not stand up. Not one sighting of a 'flying saucer' has survived serious scrutiny.

The UFO believers boast that 'only 95%' of sightings turn out to be hoaxes, or the Moon, or some other natural phenomenon. But the remaining 5% always turn out to be hoaxes or mistakes too. The UFO phenomenon arose in the late 1940s – a time of unprecedented nervousness about nuclear technology, political instability and East–West confrontation. When the world goes through one of its lurches out of kilter, people tend to look to mysteries and conspiracies as a way of making sense of it all.

It is, of course, impossible to *prove* that there are no UFOs, just as it is impossible to prove that there is no face on Mars, or to convince the doubters that the *Apollo* astronauts really did go to the Moon. But a couple of salient anecdotes ought to make even the most die-hard UFO-phile reconsider, at least for a moment.

The first accepted sighting of a flying saucer took place on 24 June 1947 over the Cascade Mountains east of Seattle. Pilot Kenneth Arnold, by all accounts a sane and sober man, was perturbed, to put it mildly, to witness a flock of strange silver things whisking through the air 'like a saucer if you skipped it across the water'. He described the unidentified flying objects as being possessed of uncanny speed and having an ability to turn on a dime. And he was very, very clear on one thing: the shape of the things he saw. Like a bat, he said: delta-shaped with triangular wings and no discernible fuselage or tailfin.

The report caused a sensation. Newspapers focused on Arnold's description of how the objects flew far more than on what they looked like. His 'saucer skipping over the water' line caught the imagination; it was a short step to the objects becoming popularly known as 'flying saucers'.

Kenneth Arnold never claimed to have seen a flying saucer. He said he saw a bat-shaped object that flew a bit like a saucer would if you skimmed it on a pond. So what? After the newspaper misreports of Arnold's close encounter, hundreds of people started seeing strange objects flying in the sky. They were invariably saucer-shaped. This does not prove that all these people were lying or deluded, or that they experienced saucer-shaped hallucinations. It could be that one man reported bat-shaped UFOs, which inadvertently became famous as a sighting of a flying disc, and *then*, completely coincidentally, an invasion of *disc-shaped* spacecraft visited Earth from another planet/dimension/time (delete as appropriate) immediately afterwards. That might have happened, but let's face it, it didn't. There are no flying saucers.

The notorious Roswell incident is barely even worth a mention. In short, it was a weather balloon that crashed in New Mexico in July 1947 – or maybe not. It certainly wasn't a mysterious triangular flying-thingy from another world which is now being kept in a secret hangar and having its technology copied. (By whom? The Pentagon? If so, why can't F-16 fighter aircraft fly at Mach 19?) There were no hieroglyphics, no strange alloys, no half-dead aliens, no autopsies. Roswell is, in a word, nonsense. It is so obviously refutable it is dull. Much more interesting is Britain's very own Roswell: Rendlesham.

In the small hours of 27 December 1980, something extremely odd happened in the depths of Rendlesham Forest in Suffolk, home to an American airbase. That foggy night, a small group of airmen saw something that for decades remained totally unexplained – one of the 5%, in other words. It appeared to be a true close encounter, not just of the first kind but allegedly of the second too: physical evidence was found. Two bored and tired military policemen, John Burroughs and Bud Steffans, were guarding the rear entrance to the base when they saw strange, flashing lights in the sky. A veritable lightshow pierced its way through the fog, they reported – flashing blues, reds and greens, accompanied by a terrifying, wailing electronic noise. Remember, this was 1980. Stephen Spielberg's 1977 UFO blockbuster *Close Encounters of the Third Kind* was still fresh in everyone's minds. What was reputed to have happened in the sky looked very much like the arrival of the mothership in Spielberg's film, right down to the almost random fairground colours and synthesized din.

Burroughs and Steffans were, quite naturally, perturbed. This could be a flying saucer; it could just as easily be a helicopter or plane in trouble, they reasoned. They radioed the base and, it was reported, around a dozen colleagues came running out into the woods.

From this point, accounts differ. Lt. Col. Charles Halt, the base's deputy commander at the time, said his men saw 'a strange glowing object in the forest ... described as being metallic in appearance and triangular in shape, approximately two to three metres across the base and two metres high. It illuminated the entire forest with a white light. The object itself had a pulsating red light on top and a bank of blue lights underneath ... it was hovering or on legs. As the patrolmen approached, it manoeuvred through the trees and disappeared'. Halt's report, innocuously entitled 'Unexplained Lights', has since become a sacred text of UFOlogy.

Of course, there could be a perfectly good explanation. 'Teasers', as Ford Prefect explains, 'are usually rich kids with nothing to do. They cruise around looking for planets which haven't made interstellar contact ... find some isolated spot with very few people around, then land right by some poor unsuspecting soul whom no one's ever going to believe and then strut up and down in front of him wearing silly antennae on their head and making *beep beep* noises.'

A lieutenant colonel at a US Air Force Base in Cold War Britain is not Teaser-fodder. Halt was not drunk or mad, nor was he a hippy-dippy Earth child in touch with his inner Gaia. He was not, in other words, the sort of person who normally sees flying saucers. I have never met the man, but the words 'crew cut', 'no bullshit' and 'alien spacecraft crap' come to mind. So his report is important.

He focused on the testimony given by three witnesses: Burroughs plus Jim Penniston and Ed Cabansag. All claimed to have seen a 'metallic craft', complete with 'hieroglyphic markings', descend into the woods. Today, Penniston (now a human resources manager) maintains that he even managed to touch the thing. 'This was a craft of unknown origin', he says. 'Triangular. My assessment was that it was not occupied'. The lights returned the following night. This time, Lt Col Halt saw them for himself.

Now the tale starts to get murkier still, with talk of Geiger counters, high radiation levels and mysterious depressions on the ground. All this was written in a series of official memos, and there was a British Ministry of Defence investigation, which contained all the reports from the American airmen and concluded that whatever had happened at Rendlesham did not pose any direct threat to the realm. The MoD report acknowledged that an unusual event had occurred but drew no conclusions as to its nature.

Word soon leaked out. It is a tenet of UFOlogy that 'the authorities' quickly do everything in their power to cover up any contact between extraterrestrials and earthlings. The fact that the most famous sightings, even on secretive US airbases, usually become public property within days if not hours is not addressed by believers. Anyway, Rendlesham became a hive of UFO activity. Articles appeared in the newspapers. Books were written. For 23 years, Rendlesham acquired the highest status in UFO circles. There was talk of parallel dimensions, intergalactic stargates and disturbances in the psychic field; eddies in the space–time continuum, even.

Then in 2003 along came Kevin Conde. He revealed that the UFO at Rendlesham was not, as was commonly supposed, an intergalactic spacecraft traversing the void at speeds far beyond light's 1000 million kph. Nor was it a portal to another universe. Still less was the craft a time machine connecting Thatcher's Britain with an era when technology will be so advanced that, in Arthur C. Clarke's words, it has become indistinguishable from magic. No, the UFO seen by the airmen was a 1979 Plymouth Volare, top speed a cool 150 kph. General Motors' finest had made a passable imitation of the starship *Enterprise*.

Mr Conde, you see, also worked at the Rendlesham base in 1980. He was a military policeman with the task of guarding the perimeter. A job which, as you can imagine, was dull. Nothing much happens of a Suffolk night at the best of times, and certainly not in 1980. So Conde decided to have some fun. He stuck some green and red tape over the headlights of his car. He messed around with the flashing lights on its roof. He tinkered with the PA system until he made it whoop and hoot with feedback, and waited for the sort of foggy night when even the sharpest eyes play tricks. 'There was this one guy at the back gate, and he was known as a bit of a problem', Conde told me from his home in Sacramento, where he now works in IT. 'He was always seeing things. He had seen lights before and reported them – it always turned out that it was a star or something. So I decided to play a practical joke. I had no idea what I had started by doing this.'

'I drove down the taxiway in my car. I stuck the spotlight on, after sticking red and green lenses on it. I then drove round in circles, in the fog, with the PA loudspeaker going, flashing my lights. It was just a practical joke; we were always playing practical jokes. Then I turned my lights off and drove away.'

Rendlesham mystery solved. The UFO was a police car in drag. Conde returned to his homeland and forgot all about it, until he looked up his old base on a US military website in 1992. He was flabbergasted. 'I was amazed. I had no idea about all this nonsense. Logic says that if life has evolved on Earth, it must have evolved elsewhere. But logic also says that if it has, it does not go around in flying saucers and land outside a quiet military base in the English countryside. I hate to be cynical, but when I see people making money out of this I have to ask myself if they are not nuts, what are they?'

Keen to see what isn't there. Memories are notoriously fickle. It is easy to retrospectively insert all sorts of imagined events to correlate with a later-learned narrative. This could explain the apparitions of three-legged spacecraft seen by some Rendlesham personnel.

Even if we knew for sure that there was no one at all 'out there', 'eyewitnesses' would still be 'seeing' flying saucers now that the concept has been dreamt up. An alarming number of people – three million of them in the USA by one oft-quoted if suspect estimate – think that at some time in their lives they have been abducted by teams of extraterrestrials who drugged them, took them off to their spacecraft and then proceeded to do unspeakable things to their bottoms. Most of these alleged victims forget what was done to them and their bottoms (typically, abductees are female; in the smaller number of male cases the aliens generally take on the form of an easy-on-the-eye human female). Only later in life, through flashbacks and hypnosis, are 'memories' of these 'events' 'recovered'. It is all very odd, and the more you read about alien abduction stories the more you fear for the future of humanity. But there you are: life's rich pageant and so on.

In 2003, Stephen Webb, a physicist at Britain's Open University, wrote a wonderful book detailing 50 possible solutions to the Fermi Paradox. After dismissing UFOs, he rattles through the usual suspects. There is the galactic quarantine theory, or 'zoo scenario', which suggests that we have not been visited because some sort of ecological preservation order has been slapped on us. But when we think of those lost tribes of the Amazon and New Guinea and all the efforts we make to stop them coming into contact with *Big Brother* and diseases and how often these efforts fail, it is hard to imagine a galaxy teeming with life obeying rules, however strictly enforced, for eternity. And why would they bother keeping their existence secret from us? After all, we do not hide ourselves from rats and termites. No, I don't buy it, and neither does Webb.

Several solutions flow from the 'planetarium hypothesis' – that the Universe is an illusion created for us by intelligent beings. There are no aliens because these beings – one would have to call them gods – chose not to put any in. Zaphod Beeblebrox experiences a planetarium when forced to enter the Total Perspective Vortex (see Chapter 11). Why would god-like entities do this? Because they can. Webb rejects this idea because of its untestability.

Another solution to the Fermi Paradox is that intelligent life is doomed. As soon as sentience meets technology, this argument goes, the writing is on the wall for the big-brained guys. It would have been all too easy for us to have put an end to our civilization, even our species, in the past few decades, by the enthusiastic prosecution of a nuclear war. That it didn't happen is down more to luck than good judgement, and we are not out of the woods yet. But the idea that ET always blows himself up shortly after splitting the atom is depressing and unconvincing. The kind of aggression that characterizes human interactions cannot be axiomatic in an intelligent species, surely?

More likely is that intelligent life eventually gets round to building its successor, machine intelligence, which tends not to be of a space-faring persuasion. Deep Thought-like computers with brains the size of planets simply sit there pondering monumental thoughts from here to eternity. All this obsession with flitting around the cosmos could be just an adolescent fad that all intelligent races simply grow out of.

Computerized couch potatoes may not arise, though, while machines remain our servants. During this halcyon period, the entire Milky Way could be explored, using plausible rocket-type propulsion technology, in just a few million years. Fleets of automatic, self-replicating machines could be sent to establish a bridgehead on a planet, explore it, send back data to Earth and then build several new versions of themselves from the raw materials at hand and launch these on new missions.

Another possibility is that the Universe is teeming with intelligent life, with itchy feet, that hasn't managed to get here yet. Stars are stupendously far apart. At a departure speed of 40,000 kph, it took the *Apollo* astronauts a couple of days to get to the Moon; at the same speed, they would have been dead for tens of thousands of years by the time they got to the nearest star. To picture the vastness of space, it is customary to instruct the reader to think in terms of billiard balls orbiting Christmas puddings around London's South Circular road. So, not to disappoint, imagine the Sun is a grapefruit atop Nelson's Column, and we are a shotgun pellet whizzing past the gift shop in the National Gallery. In this case the nearest star, Proxima Centauri, is on the beach in Crete and the centre of the Galaxy is half-way to Mars.

Nonetheless, most engineers and scientists believe that getting to the stars is do-able, if tricky. If we threw everything at the problem, we could probably build some sort of starship in a few decades. It would be crude and terribly expensive but it would probably work. Back in the 1960s, NASA looked into building spacecraft powered by an atomic bomb. The Orion project was quietly shelved when someone built a wooden mock-up bristling with guns. President Kennedy took one look at the proposed Death Star and, appalled, cancelled the project on the spot. If Orion could have got us to the stars – in theory – 40 years ago, there is no reason to suppose that alien civilizations which may be 10,000 years or several million years more advanced than ours could not do the same at the flick of a switch.

The favoured solution to the Fermi Paradox is that we have seen no evidence of alien life because there isn't any. Or that there is very little – and it is so staggeringly far away that we will never meet it, foggy night or no. Stephen Webb concludes that there are no aliens. There is plenty of life out there, he argues, but not the sort that builds spacecraft and radio telescopes.

The geologist Peter Ward and astronomer Donald Brownlee contend that, far from being an ordinary planet, Earth is a very special place. In their book *Rare Earth* they argue that an almost unique set of circumstances conspired, after a long time, to allow the evolution of highly intelligent life forms (us) that can contemplate the existence of alien civilizations and interstellar travel.

For a start, our planet formed around a star rich in what astronomers call 'metals', meaning 'any element with a nucleus heavier than hydrogen or helium'. Earth formed in a safe, stable region of the Galaxy, far from any marauding black holes, and in a sufficiently empty part of space to avoid the worst effects of local supernovae. Some parts of our Galaxy –

some parts of all galaxies – do not contain star systems conducive to the formation of life. The Milky Way's habitable zone is a thin disc of stars about 1,200 light years thick. This zone has a long history of star formation, with many recyclings of materials from earlier suns. The production of heavy elements – 'metals' – in stars and their consequent scattering during stellar explosions is the source of the Universe's useful materials for planet- and life-building. The habitable zone of our Galaxy boasts about three-quarters of its stars.

Our Sun is a single hydrogen-burning star, stable enough to last for more than eight billion years. Earth is big enough to hold on to an atmosphere and small enough to let go of much of its primordial hydrogen, unlike like the grasping gas giants. It has sufficient surface water to make oceans and support an aqueous biochemistry and, perhaps most importantly, to allow plate tectonics. The two most Earth-like planets in our Solar System, Mars and Venus, are both sterile as far as we know, possibly for lack of the climate-stabilizing effect of surface recycling, uplift and erosion. Mars's internal heat leaked away too quickly to shatter its crust into plates. On Venus, the lack of water has lead to spasmodic global resurfacings – not good for life.

And another thing. Earth does not have too *much* H_2O. Life might well evolve in a water world, but probably not space-faring life. Dolphins may be bright, but they'll find it tricky – though maybe not impossible – to get into space without being able to handle a spanner (but a super-intelligent octopus may have fewer problems).

Our luck continues. Thanks to our hefty goalkeepers Jupiter and Saturn and their deft gravitational saves, Earth has been hit comparatively rarely by comets and giant meteorites. Had we been struck more often, life, had it arisen, would never have evolved beyond the microbe stage. Most importantly, since animals and land plants evolved 600 million years ago, we have not had a hit that could sterilize the biosphere. There is some evidence that the age of the dinosaurs was brought to an abrupt end 65 million years ago when an errant piece of space debris 10 kilometres wide smashed into what is now Mexico at 50,000 kph, but even this cataclysm did little to dent life's hold on the planet.

Most remarkably of all, we have a massive moon. Luna, our satellite, is unique among the terrestrial planets in our Solar System. Mercury and Venus have no satellites; Mars's Phobos and Deimos are just pebbles. Only Pluto's Charon is a relative heavyweight, and Pluto is a frozen dead loss as far as life is concerned (probably). Our Moon's tidal pull stabilizes the Earth's spin; without it, our climate would oscillate dangerously between ice age and flaming heat as the planet wobbled round the Sun.

All this adds up, say Ward and Brownlee, to a rather special planet, and a somewhat empty Universe, in which animal life at least is extremely – perhaps vanishingly – rare. And remember, although there has been animal life on Earth for a respectable one-seventh or so of its history, there has only been intelligent animal life for the last 150,000 years – just one 24,000th of the time Earth has been in existence. Humans have probably been taking an interest in the stars for only a few tens of thousands of years and only in the last century or so, one 45 millionth of the planet's history, have we perplexed ourselves with the existence of other beings possibly orbiting those stars, and found the ways and means to travel into space, and transmit and receive radio signals.

Having said all that, in our Solar System alone there are now thought to be several places where simple biota at least could have evolved. Richard Taylor of the British Interplanetary

Society even cites our Moon as a place where life may lurk – not little Selenites perhaps, but bacteria, multiplying in the possibly damp pore spaces deep underground.

Mars's subterranean rocks may harbour bacteria-like organisms. Jupiter's moon Europa is covered with a 25 km-thick water ice crust thought to shield a gigantic ocean of brine. Discovered by the Galileo spacecraft in the late 1990s (confirming inferences made when the Voyager probes swung by a generation before), the Europan Ocean is one of the best candidates for extraterrestrial life that we know. Then there is Saturn's moon Titan. The surface of this planet-sized moon seems to be a mixture of water ice and hydrocarbon seas, rivers and lakes. Its atmosphere is a deep-frozen analogue of Earth's four billion years ago, replete with complex organic chemicals. It is unlikely that Titan's putative lakes or seas throng with methane-metabolizing monsters, but it is not impossible.

Still, our chauvinism about life, as Carl Sagan put it, continues unabated. Many science fiction writers have assumed that aliens will look like us. But it ain't necessarily so. Yes, evolution has seen several very similar body plans arise at different times from very different starting material. The best example is probably the startling resemblance between the modern dolphin and the streamlined aquatic reptile of the dinosaur era, the Mesozoic Ichthyosaur. But they look the same because their bodies do the same job – swim fast and hunt fish. It is possible that intelligent life elsewhere will be built on a vaguely familiar body plan, but it is easy to imagine all sorts of possibilities – wheeled life forms, tentacles and so on.

Another chauvinism is that intelligent life must evolve on land. Again, there is no reason why this should be. Indeed, here on Earth (just as in the *Guide*) the second most intelligent species alive is probably the bottle-nosed dolphin. One could argue that marine existence is not conducive to developing tool-using intelligence – the sort that can build radio telescopes and speculate on the presence of aliens. Astronomy might take a long time to get going for even the most intelligent ocean species. But it cannot be discounted. After all, we had large land animals on Earth for hundreds of millions of years before one of them picked up a bone and a piece of flint and set the ball rolling towards New York, world wars and the BBC.

We also assume that life must be surface-based. Yet autonomous, energy-using self-replicating systems could easily arise in all sorts of places other than planet surfaces. Sagan suggested we might find life, intelligent life even, *inside* planets. In 1976 he argued that the upper layers of Jupiter's atmosphere might house three types of organism: 'sinkers', 'floaters' and 'hunters'. The sinkers would be like terrestrial plankton, ammonia-based free-flying creatures; the floaters roughly equivalent to fish; and the hunters huge, predatory organisms many kilometres across.

And if you think the idea that creatures could live in Jupiter sounds silly, get this: physicist Fred Adams, of the University of Michigan, has proposed that, in the Universe's distant future, life forms may arise based on clusters of black holes, dark matter and strange quantum entities. 'Life' may be vanishingly rare, or it may be ubiquitous.

Not content with assuming that life will both look like us and live how we like to live, we also seem to assume that it is going to communicate as we do. The SETI project – the Search for Extraterrestrial Intelligence, based in Mountain View, California – scans the skies for radio signals showing signs of intelligence. Some scientists think we should be looking elsewhere. Arthur C. Clarke suggested in his story *The Sentinel* (the basis for the 1968 movie *2001: A Space Odyssey*) that alien intelligences might instead leave behind artefacts – an enigmatic monolith, say. In Carl Sagan's 1984 novel *Contact*, an alien message is found buried in transcendental numbers. Perhaps the strangest suggestion comes from the astrophysicist Paul Davies. In 2003 he speculated in *New Scientist* that aliens could have left their trace in the data stream that is our genetic code. More prosaically, it has been suggested, as early as 1961, by Berkeley physicist Charles Townes, that ET could simply use high-powered beams of light – lasers – to send their message out into the cosmos. We could be listening with the wrong sort of ears.

There has been a sea change in the attitude of the scientific community to the idea of extraterrestrials in the past 30 years. In the 1970s, talking about aliens was seriously *infra dig*, especially in the corridors of NASA and other such august bodies. The results from the first space probes, particularly the 1960s *Mariner* Mars missions, which found the Red Planet a dusty, airless and probably lifeless sphere, shocked the planetary community. Before that, most scientists assumed that there were lichens or mosses on Mars. The idea that this supposedly Earthlike world was home to, at best, bacteria seemed ludicrous.

The post-*Mariner* sniffiness about aliens was an attempt to distance 'proper' science from the Lowellian dream. Percival Lowell was a Boston millionaire and passionate amateur astronomer. His sighting of Martian canals in the 1890s arguably started the whole alien paradigm. Night after night he sat atop his Arizonan eyrie peering at the Red Planet through a specially commissioned Clark refractor. Although Mars would have appeared to him as having only the diameter of a dime held at arm's length, he managed to 'see' a network of razor-straight lines across its surface – lines which he interpreted as canals, built by intelligent Martians intent on bringing irrigating water from the humid polar regions to the arid tropics. Most people took his word for it, but as the 20th century progressed and telescopes got bigger and better, the 'canals' soaked into the sands of Mars. By 1930, few serious astronomers bought Lowell's thesis, although as late as the 1960s some were still prepared to defend the idea of canals on the fourth planet. But it took the NASA Mars missions to relegate him to the role of science fiction writer. When *Mariner 4* whizzed past Mars in 1965 and sent back a few fuzzy pictures of craters and dust, scientists who saw the results thought 'there goes the space programme'. With no life – no chance of it even – surely the public would lose interest in the whole enterprise? But the spirit of Percy Lowell never really went away. As Richard Taylor puts it, 'it is probably thanks to Lowell and his ideas about Mars that NASA did not cancel the *Mariner* program after *Mariner 4*'.

Now the dream is back with a vengeance. In August 1996, US President Bill Clinton took the stage to announce the finding of fossil bacteria in a meteorite known to have come from the Red Planet. NASA now has an Astrobiology Institute tasked with investigating the possibility of Martian bacteria, microbes in Venusian cloudtops, and exotic beasties in the murky depths of the Europan ocean and possibly even Titan's hydrocarbon seas. Twenty years ago, that an august journal like *Nature* would carry a story about 'communicating with ET' seemed as likely as it running a piece about *Scooby Doo*. In September 2004, however, *Nature* published a research article entitled 'Inscribed matter as an energy-efficient means of communication with an extra-terrestrial civilization'. New Jersey engineers Christopher Rose and Gregory Wright mooted that if you are not worried about time (and in a Universe which may last a googol years or more, who is?) then there may be better ways to get the message across than radio waves.

Why not send 'inscribed matter' instead of electromagnetic beams, they ask? Radio or light waves are fast, but are a highly inefficient way of communicating any message except 'we are here'. As distance increases, the power of the signal that must be sent increases by a factor of a square of the distance.

We have attempted to send an 'incribed message' already. The *Voyager* space probes carried several artefacts including a gold-plated phonograph disc, pictures of humans, allusions to physical constants and maps of the Solar System and nearby stars. We can now store hundreds of times the amount of information per kilogram of payload than in the 1970s. But we should probably keep it simple, given that until quite recently getting even earthly computers of different makes to communicate with each other was ridiculously difficult. A neat idea, suggest Rose and Wright, is to use a scanning tunnelling microscope to inscribe information on metal alloys at the nano-scale. Around a trillion bits of useful information can be recorded on one square inch of material using individual xenon atoms to write on a nickel pad. With a relatively small amount of energy we could construct and send numerous such 'messages in a bottle' – each containing substantial amounts of information about us and our doings – and spray them in all directions on simple chemical rockets. Of course, we will be long dead by the time they get anywhere, but in the very long term this might just be the most cost-effective way to talk to the cosmos.

Rose and Wright even suggest where we should look for such messages. Our best bets are 'orbits of long-term gravitational stability, or on the surface of objects in such orbits', write the duo. Our Solar System has many spots where gravitational forces effectively cancel each other out. These are called Lagrange points; things placed in them tend to stay put for millions upon millions of years. In addition 'the surfaces of various bodies on the inner Solar System are also possibilities ... our results suggest that carefully searching our own planetary backyard may be as likely to reveal evidence of extraterrestrial civilizations as studying distant stars through telescopes'.

It would be peculiar indeed if life on Earth was a one-off. Awe-inspiring. Slightly suspicious. We know life got started here and Earth is not blessed with any particularly exotic chemistry. The *Rare Earth* hypothesis does not preclude aliens, just intelligent ones. The galaxy may well be full of slime moulds, bacteria, the odd shrub and, rarely, haddock. Carl Sagan thought there might be a million or so technology-using intelligences in our galaxy. Ward and Brownlee put the number closer to one.

Will we ever know? If 'they' make direct contact: yes. Seth Shostak, one of the big cheeses at SETI, predicts that as computers' sky-scanning improves, we will find any communicating civilizations by 2025. There have been some promising signals, all one-offs; none looked intelligent. The search continues, using the giant fixed radio telescope at Arecibo in Puerto Rico and also Jodrell Bank in the UK. There are plans to co-opt much more powerful telescopes, such as the Allen Telescope Array under construction at the Hat Creek Observatory in the USA. The ATA will be smaller than Arecibo but will scan a much wider range of frequencies and will be largely dedicated to SETI.

Perhaps the best chance of discovering alien life directly will come in the next half a century or so. America and Europe have firm, costed plans to build 'terrestrial planet-finders', great fleets of telescopes in space or on the far side of the Moon, designed to spot Earth-sized planets orbiting other stars. More modest are the planned extremely large telescopes on Earth; these would detect, spectroscopically, the tell-tale signs of life in extrasolar planetary atmospheres, water vapour, oxygen and methane. A really big telescope could even detect the lights of alien cities.

The costs of a space-borne search for life are considerable – comparable to the budget for a manned lunar landing. The payoff would be enormous. Seeing a blue and green world orbiting another star would be a powerful blow to any remaining geocentrism in the human psyche. Finding out that life on

Earth was 'just one of those things' would be the next stage in the Copernican Revolution.

How the public would react to ET intelligence would depend on the nature of the contact. The discovery of Martian microbes would be revolutionary – for academics. But given the general ignorance about science it is questionable whether this *would* be such a big story. And it would be necessary to establish that the microbes were not related to Earthly life – the two planets have been swapping spit (bits of rock from both planets have been sent winging their way across space thanks to asteroid strikes) since the birth of the Solar System. Indeed, if we find fossil life on Mars, we may never know if it evolved independently.

A radio signal from Alpha Centauri would be far more stupendous. If it contained unmistakable signs of artificiality – say the repetition of a numerical sequence – we would know that there is some kind of intelligence out there. Information and images of the sender's planet and culture would send UN officials groping for their vague First Contact protocols. Millions would refuse to believe it, of course – but then there are plenty of people who still have difficulty with evolution and a round Earth.

If aliens are out there, they may already know about us. We have been broadcasting into space for nearly a century. Early TV transmissions from the 1950s should be just now winging their way past some good candidate stars for possible intelligent life. Any inhabitants of a planet orbiting Tau Bootis, for instance, will be getting their first episodes of the *I Love Lucy* show; it will be another three years before they learn about rock 'n' roll. And if we can build terrestrial planet-finding telescopes, then so can they.

Which could be bad news. H. G. Wells's warlike aliens are derided by today's politically correct ET-philes. People now tend to believe that when flying saucers come, it will be 'in peace', like the benevolent *Close Encounter* Stephen Spielberg imagined. Sadly, human history tells us that when technologically advanced civilizations run into less well-endowed peoples, the results are usually ugly for the guys with the spears and body paint. Our First Contact could well have a whiff of the Vogons about it. Maybe we should keep our heads down till we learn more.

3
deep thought

The Great Hyperlobic Omni-Cognate Neutron Wrangler could talk all four legs off an Arcturan MegaDonkey – but only I could persuade it to go for a walk afterwards.

Deep Thought

Under law, the Quest for Ultimate Truth is quite clearly the inalienable prerogative of your working thinkers. Any bloody machine goes and actually *finds* it and we're straight out of a job, aren't we?

Majikthise, representative of the Amalgamated Union of Philosophers, Sages, Luminaries and Other Thinking Persons

In the 1970s all the most powerful computers looked very much like Deep Thought. They were big and extremely expensive. They had butch names like UNIVAC and Illiac. They came in sinister black cases and had lots of winking red lights. They were usually buried deep in concrete basements under menacing buildings and did things like calculate missile trajectories and try to work out whether it was going to rain tomorrow. No one knew how to work them. No one who didn't have mad hair and a degree from MIT, anyway. Programming these machines usually involved, at best, some rigmarole with punched cards; at worst, a soldering iron. Science fiction writers of the time speculated on a future of intelligent machines running the economy, calculating the Ultimate Answer and even falling in love. At the time no machine could make a decent cup of tea, let alone serve it to you with a smile.

The mice of the *Hitchhiker's Guide* want the answer to Life, the Universe and Everything. Living in another dimension they might be, but mouse philosophy is no more up to the task than our own feeble efforts. Philosophy has all the interesting questions. They usually begin with 'whys' rather than the 'whats' and 'hows' beloved of, say, industrial chemists. But, like psychology, philosophy has so far proved itself almost laughably ill-equipped to answer these. So far philosophers have come up with a lot of self-referential arguments for and against the existence of God, plenty of stuff about the meaning of words, existence and essence, and lots of even flakier things about truth. So, decided the mice, better to let a machine get on with the job.

Deep Thought is the result. Industrial in scale and stupendously clever, this office-block-sized machine whirrs away for seven and a half million years calculating the Answer. When the mice find out that it is '42', they are extremely cross, and remonstrate with their silicon monument to computational excess. Deep Thought proffers a solution: construct a machine even cleverer, one whose 'operational parameters I am not worthy to calculate' to work out the Ultimate Question, the answer to which is '42'.

That computer is the Earth. The planet and life on it, including us, form part of its computational matrix. It grinds away for millennia to find the Ultimate Question. Just as it is about to deliver, Earth is destroyed by the Vogons at the behest of a consortium of high-powered psychiatrists, forcing the mice to construct (or pay the Magratheans a lot of money to construct) Earth Mark Two. Which just goes to show: just when you think you've got the ultimate computer, there will always be a good reason to go out and buy a new one.

A universal computing engine sounds very shiny and modern. In fact, the concept of a machine programmed to follow different sets of instructions was fairly well established in the early 19th century, largely thanks to the burgeoning automated weaving industry driven by the European Industrial Revolution. Growing wealth fuelled demand for luxury products, particularly fine woven fabrics for clothes and furnishing. Weaving intricate patterns in cotton, silk and wool was a laborious, time-consuming process usually involving small boys and big, dangerous machinery. Even if the small boys were well and truly ripped off, the costs to the mill owner – food, water, bandages – were tiresome.

Then along came a French engineer, Joseph Marie Jacquard, who designed and built the Jacquard Loom. An operator could 'program' its weaving pattern with punched cards. There was no electric power, and certainly no silicon, but this was a primitive *ordinateur*. There was even software piracy. Mill owners would pinch card decks from rival mills to steal prized designs.

In the 1820s, the English mathematician Charles Babbage drew up plans for a calculating machine called a Difference Engine, which would perform complex arithmetical calculations using gearwheels and cams. Had any been built accurately they would have probably changed the world.

The first software writer is popularly supposed to be Ada Lovelace, daughter of Lord Byron. Rich, pretty and achingly glamorous – with a brain the size of a planet – Ada befriended Babbage in the 1830s and wrote a series of algorithms for his Analytical Engine (the proposed successor to the Difference Engine). In an age when women were supposed to look nice and worry about frocks (if they weren't down the pit or minding sheep) she was a mathematical genius. Lovelace was not the first to think about programming – Babbage knew what he was doing, as did his three sons and one of his assistants. But she was probably one of about half a dozen people in the 1830s who could see where the technology was going. And she told the world about Babbage's engines, writing the best contemporary account of a man who was probably at least a century ahead of his time.

Unfortunately Babbage's engines didn't work. There was nothing badly wrong with their design. Like Leonardo da Vinci's flying machines, they would have worked had anyone managed to build one properly (as has now been shown for both the Difference Engine and some of the designs of the ultimate Renaissance Man). It was simply that early Victorian engineering tolerances were not quite good enough for the intricate assemblies required. It is entertaining to speculate what would have happened had Babbage's blueprints been turned into reality in the 1830s. What effect would a fully functional, programmable mechanical computer have had on Britain's nascent Industrial Revolution?

The true dawn of the computer age came in the 1940s with the development of an electronic difference engine, with Babbage's brass components replaced by valves and electrical circuits. Like so much of the modern world, computing was forged in the fires of the Second World War. Both the Allies and the Axis powers realized that universal calculating machines would be of tremendous use. The most famous computing project in the war took place at Bletchley Park, where some of the finest minds in the world (including computing pioneer Alan Turing) laboured to break the codes used by the German High Command. As the war progressed, the process became more and more automated, and eventually Colossus, a semi-programmable computer, was built to crack the most fearsome codes, the Lorenz cyphers. By the 1950s, computer engineers were beginning to sketch machines that could tackle fiendishly hard sums, perhaps the hardest sums of all: weather forecasts.

As early as the 19th century British forecasters with the Royal Navy attempted to make sense of the limited data gathered by ships and on-shore weather stations. In the 1890s, *The Times* was brave enough to run summaries of these forecasts. They were sometimes right, but more often hopelessly wrong. The notion that you could forecast the weather with any accuracy fell out of fashion in the early 20th century, especially when, in the 1920s, British mathematician Lewis Fry Richardson drew up the first proper equations to predict the movement of air masses and fronts. He also calculated what it would take to plug the figures from weather stations into these equations. His conclusion? 64,000 skilled and trained weather forecasters working flat out in a single building should do it. Even with this unlikely workforce, tomorrow's weather forecast wouldn't actually be ready until two days later. What was needed was a computing machine to do the tedious pencil work. And so it came to pass.

After the War, the British Meteorological Office began using Leo, the computer that ran the then vast Lyons Tea House chain. For a short while, Britain's sailors, farmers and tourists relied on forecasts calculated by a machine which spent most of its time counting sticky buns, fruitcake and clotted cream. And for the next couple of decades, that is where computing stood. Big machines did impossible number-crunching jobs for the men from the ministry, tended by chaps in flat caps and blue uniforms equipped with soldering irons and oil.

Then along came NASA. The computer power available to the US Space Agency at the beginning of the 1960s, when JFK pledged to put a man on the Moon before the end of the decade, was about the same as that of a microwave oven today. NASA powered forward developments in transistors and microprocessors – the now-ubiquitous silicon chips – driven by tens of billions of dollars of taxpayers' cash. When Buzz Aldrin and Neil Armstrong guided the *Eagle* lunar lander down towards the Sea of Tranquility in June 1969, their flimsy little craft was equipped with a compact, powerful little computer that ran the guidance systems, controlled the fuel supply and sent helpful messages to the astronauts telling them so. It was hand-held, and had an LED display screen and a simple keypad. It was no Deep Thought – it didn't stand a chance of coming up with the answer to Life, the Universe and Everything – but the *Apollo* computer was the true first wave of the modern silicon revolution.

There are now about half a dozen computers in my car, each far more powerful than the machines that guided Armstrong and Aldrin down. The elderly but hard-working PC I write on has more processing power than all the world's computers put together when the Leo was forecasting the nation's weather. And there are millions of these things, all over the world, running spreadsheets, sending smutty emails, searching for secondhand cars, and carrying out Nobel prize-winning calculations.

Even twenty years ago that this was going to happen was not obvious. In 1981, Bill Gates, the $50 billion boss of Microsoft, allegedly said '640K [of computer memory] ought to be enough for anybody'. Most home computers now have around a thousand times that amount in RAM, and half a million times as much again on their hard drive. Similarly, future IBM Chairman Thomas Watson is supposed to have said 'I think there is a world market for maybe five computers' (to be fair, this was in 1943).

One man whose predictions have stood the test of time is Gordon Moore. In 1965 the founder of silicon chip behemoth Intel famously predicted that the power of top-end computers would roughly double every 18 months. What he actually forecast was that the number of transistors on an integrated circuit would double every 18 months, which amounts to more or less the same thing. So far, his prediction has held. My partner's computer (newer than my trusty old PC), made in 2003, runs on a 3 gigahertz Pentium 4; it has half a gigabyte of RAM and about 100 gigabytes of hard drive space. My first PC, bought in 1995, was about 60 times slower and had one fiftieth of the memory.

My PC can play some wonderfully gruesome games, but not by a long way could it talk the hind legs off an Arcturan MegaDonkey. It couldn't work out the Ultimate Answer to Life, the Universe and Everything, even if I left it running for six million years (a laughable prospect since it runs on a certain popular operating system that renders it bound to crash after less than 6 million seconds). It can perform several million calculations a second, but it's no brighter than a cockroach.

At the time of writing the fastest machine in the world is a device with the catchy name of 'BlueGene/L'. Built by IBM for the US Department of Energy's Lawrence Livermore National Laboratory in California, BlueGene/L runs at 70.72 teraflops. One teraflop is equal to one trillion calculations a second, the standard unit of supercomputer speed. The second-fastest computer, NASA's Columbia, runs at 51.87 teraflops, and in third place Japan's NEC Earth Simulator runs at half the speed of BlueGene/L; it can do a paltry 35 trillion calculations a second. Like new parents, BlueGene/L's creators boast that their baby will soon be off its hands and knees to trot at an even more impressive 250 teraflops. Compare this to the world's first supercomputer, the Cray-1, turned on in 1976 at Los Alamos, which could manage a then-staggering 80 megaflops. The Cray-1 may have looked the part (big, black, sinister, plenty of flashing red lights), but compared to BlueGene/L it was a mere abacus; a fossil, a relic of the past as primitive as the ancient timepieces on display at London's Maritime Museum. It was half a million times slower.

BlueGene/L is no 42-monkey. It models the safety and effi-
cacy of America's nuclear arsenal. This has become something
of the career of choice for the world's fastest computers once it
was realized that actually exploding atomic bombs above or
below ground was (a) very dangerous, (b) gets you into trouble with pesky environmentalists and (c) violates international
treaties anyway. Some nations of course were until quite
recently happy to blow up the odd South Pacific island as part
of their nuclear testing programme, but nowadays most desist
from this sort of naughtiness in favour of computer modelling.

Computers several thousand times faster again than
BlueGene/L should be built before we reach the physical and
practical limits of the gadgetry that runs today's top-end
machines. Then, engineers warn, Moore's Law will be toast.
You can only squeeze so many circuits onto a silicon wafer
before connections have to be less than an atom thick, and as
you try to squeeze more out of a chip the electricity being
pumped in raises temperatures to the point where the circuits
start to melt.

What if Moore's Law held for centuries? In 2002, Seth Lloyd,
a Professor of Mechanical Engineering at MIT, wrote a lighthearted piece on just this question for *Edge*, an on-line journal
of cutting-edge techno-musings.

> ... in about 600 years ... the whole universe will be running
> Windows 2540, or something like that. 99.99% of the
> energy of the universe will have been enlisted by Microsoft
> by that point, and they'll want more! They really will have
> to start writing efficient software, by gum. They can't rely
> on Moore's Law to save their butts any longer.

The saviour of Moore's Law could be quantum computing.
Proponents hope to perform parallel calculations exponentially
faster than is possible with silicon processing by harnessing
weird quantum phenomena – such as the way that electrons

can be in several places at once or eerily connected to other faraway electrons. Quantum computing promises much – ultra-fast processing, the creation of unbreakable codes – but faces formidable hurdles. To date, no one has managed to build a quantum machine that can handle more than a few bits at a time.

The 64-teraflop question is: could a Universe-sized computer harnessing the potential power of every atom actually *think* or are thinking machines impossible? In 1978 you would have got long odds on a computer ever being built that could even challenge the world's greatest chessplayers. People assumed that to beat a Grandmaster would require the shadowy, elusive qualities that separate humans from mere abacuses – qualities like hunch, intuition and flair.

Chess is a fair test of human vs. machine, something Arthur C. Clarke got right (along with so much else). In *2001: A Space Odyssey*, HAL, the aberrant spacecraft computer, whiled away the hours challenging the astronauts to a game or three. Chess is, in theory, number-crunchable. There is a finite quantity of possible games that can be played. Unlike poker, say, it can be described and analyzed digitally. There are 32 pieces, each with set powers. There are 64 squares onto which they can move, and a clear definition of what it is to win (or draw). You should be able to give a computer the starting position on a chessboard and it should be able to calculate every permutation of moves and come up with one of three answers: white wins, white loses, or a draw.

The problem is that there are more chess games than there are atoms in the Universe. No computer has come near calculating the problem. (Checkers or draughts, being much simpler, has actually been 'solved', so tournament players now have to start from randomized positions to prevent drawn games simply being played from memory.) Yet in 1997 IBM's Deep Blue beat world number one Garry Kasparov in a six-game match. Its triumph hinted that it might be possible to close the gap between human and machine intelligence. Deep Blue, like all computer chess masters, takes a brute-force approach to the game. It says something remarkable about human intelligence that biological judgement and only limited calculating powers are enough to match such silicon behemoths.

Conscious machines are the dream of many Artificial Intelligence researchers. Every few months someone claims that they are around the corner. Soon, they say, processors will mimic the human brain; soon we will download our very thoughts and emotions. Humans and computers will merge in a sort of cyber-driven evolutionary leap, and by the 2030s the big moral debates will surround not therapeutic cloning, drugs or euthanasia, but the rights of machines.

This is, of course, drivel. Even when uttered by reputable scientists this kind of futurology should be taken with a shovel of salt. We do not know if we can build a thinking machine because we have no idea how a thinking brain works. We know a lot about neurons and synapses, and people have mapped how various bits of our brain link up and which bits are responsible for what. But when it comes to understanding the human mind we are about as well equipped as a group of primitive Pacific Islanders contemplating a crashed aircraft. They could weigh it, describe it, measure it – take it apart even. They would have seen it fall from the sky and hence infer something of its function. As to grasping the mechanics and physics of flight? No way.

This doesn't mean that computers do not have personalities. One of the (many) things that Douglas Adams correctly predicted was the desire to engineer fake emotions into electronic devices. When Zaphod Beeblebrox instructed the elevator in the *Hitchhiker's Guide* offices to take him to a top floor, he is reminded that he should consider 'all the possibilities that down might offer'. This was a satire on the banalities of American service industry-speak, but it has come to pass. Car satellite navigation systems are now imbued with quite distinct personalities. I recently borrowed a top-of-the-range BMW and was told, in a firm, crisp and, dare one say it, Teutonic manner when to turn right, when to turn left and when one had been a naughty boy and not followed instructions. One likes to imagine the systems plumbed into Italian cars taking a much more relaxed view of things, and those sitting behind the dashboards of French conveyances instructing their drivers to find the nearest available restaurant and take a five-hour break for lunch.

Quite apart from the kind of computers that would put Majikthise and his colleagues out of a job, we have not even built mechanical slaves to relieve us of our tedious, difficult and dangerous jobs. The kind of robots that most of us associate with the word – the general-purpose humanoid slaves first imagined by Czech playwright Karel Čapek back in the 1920s – have to date been little more than toys. Roboticists tend to talk about automated systems, autonomous learning, production lines and manufacturing processes. There are all sorts of problems that no one would have thought of fifty years ago: how do you power your robot? How do you get it to interact with the physical world? To date the best efforts have been faintly comic – foiled by stairs, or running out of battery-puff after a few minutes. Marvin's brain the size of a planet may soon come to pass, but his body – complete with faulty diodes – may take a lot longer.

4

the existence of god

'I refuse to prove that I exist,' says God, 'for proof denies faith, and without faith I am nothing.'

'But,' says Man, 'the Babel fish is a dead giveaway isn't it? It could not have evolved by chance. It proves you exist, and so therefore, by your own arguments, you don't. QED.'

'Oh dear,' says God, 'I hadn't thought of that', and promptly vanishes in a puff of logic.

The Hitchhiker's Guide to the Galaxy

Forty-two may be the Ultimate Answer, but what is the Ultimate Question? Predictably enough, nobody seems to know. It may be the case that the Question and Answer are mutually exclusive. One cannot know both – not in the same universe, anyway. The machine – Earth – built by the mice is destroyed just as it is about to come up with the answer. And even if it hadn't been blown up by the toxic Vogons to make way for a hyperspatial express route, the fact that Earth's indigenous population of *Homo sapiens* was elbowed out by a bunch of castaway telephone sanitizers and management consultants in the early Pleistocene meant that getting a straight answer out of Earth was never going to be easy anyway.

Perhaps the mice could simply have asked God. After all, if He doesn't know, then the Ultimate Answer surely isn't worth a penny candle anyway. Not to mention the Ultimate Question. So do we have any evidence of God's thoughts on the matter? Well, there is, of course, the Supreme Being's (alleged) Final Message to His Creation, which takes the form of a tawdry tourist attraction in the land of Sevorbeupstry, on the planet Preliumtarn whose solar system sits somewhere in the middle of the Grey Binding Fiefdoms of Saxaquine. Getting to see the message involves a long and tiresome hike across a sunbaked desert plain. The payoff could be argued to be a tad on the disappointing side. 'We apologize', the message says, 'for the inconvenience'.

Snappy, certainly; catchy, just about. It was, we are told, good enough for Arthur, Fenchurch (who had once realized something similar) and Marvin. But, it has to be said, as an example of Divine pronouncement it is distinctly lacking in, well, inspiration. And it doesn't really have anything much to do with 42, or any other number. Indeed, you might be forgiven for suspecting that the 'message' is just something cooked up by the guys on the scooters who have the concessions. The word of God? Well, that depends. And what it depends most on is whether or not there is a God at all.

The question of the existence of God has troubled Earthly, and galactic, philosophers muchly. More than a century after Einstein had his bright idea, and getting on for 170 years since Darwin had his, a large proportion of the planet's educated population still reckons that humans and all the other species were created, a few thousand years ago, by a Middle Eastern Sun god ... in a week. Things being what they are at the moment, one might have assumed that any Middle Eastern deities would be keeping their heads down. Instead, there is alarming evidence that God has never been so popular.

Is the existence, or otherwise, of God a scientific concern? Yes. Until say, 600 years ago, 'God' was a highly satisfactory solution to the problem: 'Where did it all come from and what does it mean?'. A lot of data – the existence of the Earth, the twinkly lights in the sky, the plants and animals and so on – needed explaining.

If you mate two rabbits together, you get more rabbits. If you mate these, you get more rabbits. They do not magically turn into tortoises, or giraffes or petunias. Similarly, a pond full of fish keeps on spawning fish. It simply does not happen, as far as human experience goes, that, one day, one of the fish grows legs and crawls out and starts bothering the cat. In the days before genes and DNA, before cells and mitosis, gametes and the whole complicated panoply of post-Darwinian biology (not to mention the knowledge that the Earth is almost indescribably old), explaining where we, the rabbits, petunias and fish came from was tough. The idea of a focused creator putting in the hardest week's work in history and conjuring up the whole thing took the problem away and put it in a nice cupboard where we could forget about it.

The history of scientific rationalism versus religion has been one of humans taking item after item out of this cupboard labelled 'God', having a good look at them and putting them in their proper place – jars marked 'evolution', 'large flaming balls of hydrogen billions of kilometres away' and 'not flat, but round, definitely round'. Today, thanks to fossils, radiometric dating, Martian meteorites and so on, God's cupboard is looking rather bare. We can't yet explain the weirdness of the quantum world, the nature of consciousness or what happened before the Big Bang – but having cracked 'where do rabbits come from?' we're confident that we will do soon. No wonder the modern deity has been christened the God of the Gaps, relegated to explaining that which science has so far failed to do.

Nonetheless, our friend with the white beard is showing signs of a comeback. Fewer than a third of Americans polled are reported to believe that the Earth is billions of years old and that humans and all the other species evolved by natural processes. Nearly 80 years after the State of Tennessee made itself a laughing stock by prosecuting a young schoolmaster, John Scopes, for teaching evolution, Darwin is back in the dock. In the late 1990s a flurry of court rulings and decisions by education boards put 'intelligent design' – a politically correct version of creationism – onto the curriculum in several states on a par with natural selection. Evolution is, the god-botherers argue, 'just a theory'. So is gravity, but apples don't fall upwards and jumping out of a third-storey window is going to hurt (unless, like Arthur Dent, you manage to miss the ground).

The God of the Bible is not the only one to have crept back onto the agenda. The vacuum left by the widespread rejection of traditional worship has been filled with the flyblown faith buffet that is 'spiritualism'. We have pop stars espousing expensive sects, billionaires professing themselves Buddhists, and a general feeling that to be 'spiritual' is a Good Thing. This rise of religion-lite poses maybe a greater threat to science than the concomitant return of old-fashioned Deep South, burning-cross conservatism. The study of the natural world *per se* is not imperilled – but the principle that logic and evidence should count for more than superstition and hunch is. After all, when it comes to deciding whether modern medicine is more likely to cure you than, say, ingesting some Malaysian bark chippings and chanting, priests and popes tend to defer to doctors; the New Age brigade embrace the darkness.

But we digress. Back to God, proper God, the Creator, the omniscient and omnipotent deity revered by billions, the God of wrath and damnation and hellfire and all the rest of it. What about him (or Him, as he, sorry He, insists on being called)? What is the point of conjuring up an invisible super-being who can see your innermost thoughts and is planning what will happen to you after your funeral?

That there is no culture or people on Earth without a strong belief system in a supernatural being or beings is good evidence that religiosity is probably hardwired. Belief in a shadow world of spirits and ghosts, at least, may predate the evolution of many of the things we consider to be human, such as language and the use of fire. Several graves have been discovered in Europe containing skeletal remains of our extinct cousins *Homo neanderthalensis* along with petals and beads. Either these fell in by accident, or the knuckle-dragging ape-men of popular myth sent their colleagues off with a degree of ceremony unknown in any contemporary species save their cousins, us. Why bury the dead with flowers if you do not believe that some essence of them – their soul – will survive the experience in some way?

The civilization that flourished along the banks of the Nile from 3000 BC was extraordinarily devout. Any visitor to Egypt cannot fail to be impressed by the effort and dedication that went into constructing the amazing temples, pyramids and tombs that glorified the complex pantheon of ancient Egyptian life. This was a society on the cusp of technology, where life for most was hard and short, yet it found the time (and whatever passed for money) to raise awesome religious structures that we would have difficulty recreating today.

Truly atheist societies are extremely rare. The Communist experiment, which ended abruptly in the late 1980s, save in the redoubts of North Korea and Cuba, had state-sponsored atheism as one of its ingredients. It 'worked' in so far as worship dwindled in countries like the Soviet Union, Bulgaria and Romania – nations previously quite devout. But the shift wasn't lasting; atheism has been on the retreat since Communism's collapse. Church attendance is booming in the New Russia as never before. Even seminary-educated Stalin realized the power of religion and effectively un-banned God during the USSR's darkest hour, when the Panzers were knocking on the Moscow gates. The relative religious apathy of today's western

Europe is an anomaly, geographically and historically. Church attendance as a percentage of population is five times higher in the USA than UK.

Showing that everyone has always believed in God is not the same as showing that God exists. Many of the traditional 'proofs' for God were formulated by the 13th century philosopher and theologian St Thomas Aquinas. He came up with the concept of the prime mover – a being that causes everything in the Universe to move – and the 'nothing is caused by itself' argument. This states that everything in the Universe is caused by something else. A table is made by a carpenter, who in turn was made by his parents and so on. This cannot go back *to infinity*, reasoned Aquinas, so there must have been a first cause: God. This is actually quite a satisfactory argument. The alternative is the philosophically unnerving prospect of a Universe without beginning or end.

The problem with the first cause argument is that it falls foul of Occam's razor. William of Occam, a medieval monk, came up with one of the most useful tools in the history of thinking. Put simply, the razor shaves away unnecessary entities. We agree that the Universe seems to require a beginning. The alternative is that it has been here for an infinite amount of time, and infinity makes philosophers unhappy. If it does require a beginning, then something must have been around to bring it into existence. Hey presto, God. But, hang on a minute. We haven't really answered the question of where did everything come from. Because we are now left with a new problem: namely, who or what created God? If the answer is 'someone else', then we defy Occam's razor – we have multiplied entities, by postulating a super-creator responsible for making God. Of course, we then have to ask who created *him*, and so on.

Of course no believer in deities will answer that God had a creator. They will say that God has no beginning and no end, existing outside time. To talk about God and the Universe being the same sort of thing – i.e. entities that must obey the cause and effect rule – is to make a profound category mistake. To believers, God is exempt from cause and effect as an eternal, timeless being who always has been and always will be, a different sort of 'object' to the Universe and everything contained therein.

A less satisfactory argument for the existence of God came from René Descartes, the 17th-century French philosopher. At the time it was thought quite convincing, although reading it now one is left with the feeling that this may have been the point when French philosophy started the inexorable descent into its own fundament which led in time to the supreme pointlessness of the existentialists and deconstructionism.

Descartes' argument went something like this: I exist, I can imagine God, I am imperfect, imperfect beings cannot invent the idea of God – ergo, God must exist. He does not explain *why* imperfect being cannot invent the idea of God – after all, he would have to accept that if God did *not* exist then he, a human, could still be imperfect and yet still invent him.

Descartes' argument is one of the so-called ontological 'proofs' for the existence of God. Ontological arguments attempt to prove God not by direct observation of the world (i.e. experiment) but by reason alone. To a non-philosopher – particularly a modern non-philosopher – these ontological 'proofs' sound highly suspect indeed, yet for centuries they were considered some of the finest weapons in the theists' arsenal. It would probably be facetious to summarize the ontological argument as 'believing that there is no God is silly, therefore he exists', but this in essence is what it says.

The ontological argument was neatly demolished by (among others) Immanuel Kant, who pointed out that it relied on 'existence' being a property, or predicate, of an object just like its colour, weight and so on. Clearly, whether something exists or not is a different question from what colour it is or how much it costs in the shops, or even whether or not it is perfect.

Of the many 'proofs' posited down the ages, one is back in the spotlight. The 'argument from design' is the line the neocreationists trot out. These people do not profess to believe that the Earth was created 6000 years ago with Adam and Eve. They maintain that the Universe is so obviously set up to allow for the development of human life that the hand of a creator must be responsible.

Take the values of the physical constants. If gravity, the electroweak force or the strong nuclear force were the slightest bit different, our Universe, with its hydrogen-burning suns and small, rocky, life-friendly planets, would be impossible. Make gravity a bit stronger and the Universe would have collapsed quickly after the Big Bang. Weakening the strong force by even 5% would wreck the fusion reaction that powers stars. You only need vary each universal constant by a small amount to get a sterile Universe, composed entirely of hydrogen gas and no stars, or just black holes or cool red dwarfs, or one that blows itself to bits in a few million years. More impressive even than the physical constants are the three spatial dimensions. It is hard to imagine intelligent life evolving in a flat universe.

There are other suspicious pointers to the hand of God, this argument goes. Water, say. Water is one of the weirdest chemicals. Unusually its solid phase – ice – is less dense than its liquid phase. So when water freezes it floats. If it were like almost every other compound, Earth's oceans would be mostly solid ice, with a thin layer of liquid over the temperate and tropical zones, and life would probably never have evolved. Water's boiling point, specific heat capacity, surface tension and melting point are all anomalous. It remains liquid, for example, at a far higher temperature than standard chemistry predicts for a molecule of its molecular weight. And it is this oddity that is the stuff of life.

Any theory of the existence and origin of the Universe must allow the development of creatures like us, otherwise we would not be here pondering such theories. This is the famous Anthropic Principle, and explaining why this fine tuning should be so without recourse to creating a deity taxes many a secular scientist's mind. It is not known whether the laws of physics we observe in our Universe are the only ones possible. Maybe the strong force cannot be other than we observe it to be, or maybe gravity can work in reverse. Maybe, underpinning the laws of physics, are yet more immutable laws of mathematics and logic. If it turns out that the Universe we see is the only one that *can* be, this too would seem to support intelligent design, as it brings us back to the fine-tuning problem.

Enter the Many Worlds hypothesis. This nails the anthropic problem by postulating that our Universe is just one of an infinite number of such. If there are an infinite number of universes, all different, then its no biggie that ours is fine tuned for life. There are oodles more that are not and are hence sterile (more of which anon). Just as we only find fish in water, we only find life in universes where it can exist. That our Universe is fit for life is then no more surprising than finding a suit that fits in a clothing store. We do not need to invoke a creator at all, unless of course we are working in a physics department situated south of the Mason–Dixon line. End of problem, God shown the door.

The origin of morality is often presented as a problem for non-believers. There are many things we take to be true about the world: that stones fall to the ground if we drop them, that

Rome is the capital of Italy. And that it is wrong to cause unnecessary pain and suffering to others.

The last of these is different from the other two. It is incontrovertibly the case that Rome is indeed the capital of Italy. You can go there, see the Parliament, meet the Prime Minister and so on. You can drop stone after stone, measure their falling and deduce all sorts of things about gravity and mass. But what about that last, moral 'fact'? While ideas about property, sex, family life and the treatment of criminals vary across time and place, most people in most cultures have agreed on a few moral absolutes. Hurting people for no reason is wrong, say, as is taking property generally accepted to be another's. 'Do as you would be done by' seems to be the universal creed.

Religious folk argue that such ubiquitous high-falutin' absolutes, being antithetical to furthering one's own interests, must have come from somewhere outside our normal rule-making processes, i.e. God.

On the other hand, lots of apparent altruism can be explained in terms of pure biology. Grandparents benefit from caring for their children's offspring. They help their own genes (albeit diluted a generation) to survive and hence propagate by their action. Even more distant acts of familial service are similarly explicable. The devotion of an uncle to his sister's children is quite easy to explain: unlike his 'own' children, he feels he knows for sure that these carry his genes.

Harder to explain are acts of genuine kindness to complete strangers. What compels an adult to leap into the ocean to rescue a drowning child? What makes millions of people all over the world donate sizeable sums of money to charities that help people they will never even meet? To the religious, such acts can be evidence that a benevolent spirit guides us all. To non-believers explanations are – must be – more complex. Humans are a societal species. We behave to our fellow men in a way that keeps the wheels of society running smoothly because society helps our offspring survive to reproduce.

People with no sense of morality, such as psychopaths, may prosper in the short term, but tend to meet a sticky end, whatever all those movies glamorizing the gangster lifestyle would have you believe. Most career criminals spend most of their lives in prison, or find themselves prematurely dead. In evolutionary terms, people who do not behave at least a little altruistically are often failures, with the odd, blistering exception – the exceptionally talented psychopaths such as Hitler and Stalin, who manage to persuade others to follow the path of their burning aggression.

Robert Trivers, a sociobiologist at UC Santa Cruz, has developed the theory of 'Reciprocal Altruism'. His experiments show that what can seem like altruism usually works on the principle of 'you scratch my back, I'll scratch yours'. Our unrelated tribemates are just another selective pressure, like temperature or rainfall. If it turns out that altruistic behaviour toward them helps us survive – by generating reciprocal altruistic behaviour – then that ability may endure Darwin's mill.

Perhaps the most ingenious recent argument for the existence of God comes from a little-understood but extremely powerful branch of mathematics called Bayesian statistics. In his 2004 book *The Probability of God: A Simple Calculation That Proves the Ultimate Truth*, British physicist Stephen Unwin concludes that the chances are considerably better than evens – 67% to be precise – that God or 'Proposition G' – exists. Says Unwin:

Our Proposition G refers to the God of Christians, the Jehovah of Jews, the Allah of Muslims, the Wise Lord of Zoroastrians, etc. Although there is some disagreement between and within religions about the specific characteristics of the person-God, the similarities in beliefs outnumber the differences. Put another way, followers of these religions could be relied upon to gang up on any hapless pantheist who found herself in the wrong part of town.

Bayesian statistics is named after Thomas Bayes, an 18th century English mathematician and minister. In the 20th century this 'mathematics of uncertainty' became a powerful tool in medicine, insurance, weather forecasting and criminology. Put simply, Bayes' Theorem states that we can assess the probability of something being true or false from a series of facts and assumptions.

Unwin starts from the premise that God has a 50:50 chance either way of existing. He then factors in miracles, free will and so on to come up with a figure indicating the likelihood of there being a deity in charge. For each phenomenon, he assigns a 'Divine Indicator', a measure of how likely that it is in a godful universe than in a godless one. What this amounts to, Unwin claims, is a 'sort of scale that does for God-related evidence what Richter did for earthquakes and Fujita did for tornadoes'.

The presence and recognition of goodness in the world, for example, Unwin labels a strong indicator that the Big Man really is upstairs. Evil, on the other hand, he takes as evidence against the idea. Interestingly, supernatural occurrences, visions and so on count for the atheist prosecution, so to speak, as almost all turn out to be false.

Superficially, Unwin's argument looks pretty convincing, but the more you stare at it the foggier it becomes. Why the *a priori* assumption of a 50% chance of God, for example? The figure is the statistical default for something that either exists or does not. In practice it seems perverse, as the tooth fairy illustrates. Either tooth fairies are real or they aren't. But only a lunatic would assert that there is a 50% probability that there are indeed small winged hominids that take the shed milk teeth of infant *Homo sapiens* in return for a small donation in an appropriate currency. We have plenty of evidence that this is not the case: no one has ever seen a tooth fairy; parents always own up to putting the coin under the pillow. And there is the rather important fact that if tooth fairies were real we would have to rethink a large chunk of what we believed to be true about Life, the Universe and Everything.

Unwin counters that a Martian landing on Earth, tasked with establishing the existence, or otherwise, of the tooth fairy would *have* to start out with a 50:50 prior probability and work from there. 'In the God analysis', he says, 'I wished to keep my prior pristine and use Bayes' Theorem as the sole means of accounting for the evidence.'

In the end, no one is going to be convinced by a mathematical argument for or against God. Like the Babel fish, Unwin has probably done God a greater disservice than Darwin. Believers are unhappy that the result is a mere 2:1 in favour; they were expecting something far more conclusive. Atheists are equally scathing. 'If you put rubbish into mathematical calculations, rubbish is what will emerge', says Richard Dawkins, Oxford biologist, friend of the late Mr Adams and perhaps Britain's best-known atheist.

So it doesn't look like the existence or otherwise of God is set to be resolved anytime soon. Just as well, perhaps, because it's a question that has probably given us humans more excuse for arguing than anything else in our history. And there is nothing that we carbon-based bipeds like more than a good argument.

5

the restaurant at
the end of the universe

This ladies and gentlemen is the proverbial 'it'. After this, there is nothing. Void. Emptiness. Oblivion. Absolute nothing ... Nothing ... except of course for the sweet trolley, and a fine selection of Aldebaran liqueurs! And for once, you don't need to worry about having a hangover in the morning – because there won't *be* any more mornings!

Max Quordlepleen
MC at the Restaurant at the End of the Universe

In an allegedly infinite universe, with an infinite number of marketing men and focus groups, a restaurant where diners can witness the demise of all that exists is bound to come around sooner or later. Thus, Milliways, the Restaurant at the End of the Universe, is born. The laws of physics may be about to take their final bow, but the laws of catering – throw in some glitzy lighting and spend a lot of money getting the bar done up in lizard skins and you can charge the galaxy to credulous tourists – hold fast to the final nanosecond. If not beyond. Exiting diners must still negotiate the financial and moral complexities of the valet car park.

Like all restaurants built to show off a spectacular view (think of those rotating places atop skyscrapers or beside waterfalls), Milliways is frequented by the rich and beautiful – in this case from across the Universe and from each end of time. And like all restaurants famed for where they are rather than for who is in the kitchen, Milliways is rather ghastly. The food is nothing to write home about, apart from the bovine bred to want to be eaten. The real spectacle is outside. As the MC announces: 'photon storms gather in swirling clouds around us, preparing to tear apart the last of the red hot suns...'.

Most scientists now believe that the Universe will end relatively peacefully. Until recently, relatively little research had gone into how this will end; science was too busy probing how things began. Now, physics, astronomy and cosmology are increasingly turning their attention to the ages to come as well as those past. It is one of the quite astonishing achievements of science that we can say with some confidence what will happen to the Universe in the countless trillions of aeons to come. We cannot predict the weather a week from today, but we can forecast with far more precision what the ultimate fate of the Earth will be six billion years hence.

Paradoxically, when it comes to the near future, we simply don't know what is going to happen to the Earth. Grim predictions concern climate change, and in particular human-made global warming, which many maintain is about to cause a rapid, catastrophic and perhaps permanent shift in the world weather. Sceptics counter that most of the predictions are based on poorly tested computer models and that, anyway, the Earth's climate has swung wildly throughout geological time.

In the long term, the arguments will become moot, as whatever we do some very nasty global warming indeed is heading our way. The problem is not us and our insatiable appetite for gas-guzzling machinery. It is the Sun.

Our star is a typical hydrogen-burning object, the life cycle of which is now fairly well understood. Like most of the stars in the sky, the Sun is a 'main sequence' star. These giant fusion bombs can range from red dwarfs with a few per cent of the mass of the Sun to giants like Sirius, the brightest star to the naked eye.

The Sun is mostly hydrogen. At its centre, the gas is so compressed and heated by its own weight that nuclear fusion takes place. Individual atoms of hydrogen combine to form helium, the next element up in the Periodic Table. In the process, a modicum of mass is lost (around 0.7%). This mass is converted into an amount of energy determined by Einstein's famous equation $E = mc^2$. As c is the speed of light, c^2 is a very large number. In other words, the Sun does not have to process much hydrogen to keep itself alight. A few thousand tonnes per second keeps its core at a temperature of 16 million degrees.

The Sun has been burning hydrogen since a few million years after the Solar System coalesced out of a ball of gas and dust about 4.5 billion years ago. As a youngster it was much dimmer than it is today. Four billion years ago or so, when the first primitive life forms arose in the Earth's primordial seas, the star's output was only about three-quarters of what we see now. Over time, the amount of helium 'ash' in the solar core has increased, making the Sun grow hotter and brighter – by an imperceptible amount in a lifetime, even in that of a civilization – but by a significant amount over the aeons of geological time.

It is important to note that any global warming we are seeing now is *not* caused by this long-term brightening of the Sun. Superimposed on this steady change are thousands or millions of short-term cyclic variations in solar output, probably related to eddies and currents in the Sun's interior gases, and to changes in the Sun's magnetic field.

It is this long-term overall warming which seals the fate of the Earth. The Sun will run out of fuel in six billion years. Long before that, things will start to get ugly. In maybe as little as a billion years, Solar output could have increased so much that an irreversible climatological disaster will befall our green and blue planet. If the Sun is maybe a quarter as bright again as it is now, global temperatures could increase by 20 °C or more. This in itself is not a problem: life could adapt given the vast time-scales involved. And a steadily warming Sun may not be accompanied by a steadily warming Earth. Feedback mechanisms may kick in to ameliorate the effects. The most obvious is cloud formation – a warmer Earth means more water evaporated from the oceans and hence more clouds. More clouds mean a whiter Earth and more light reflected into space, which would act as a brake on warming.

Nevertheless, there will come a time where no amount of feedback will protect the Earth. At that fatal tipping point a runaway greenhouse effect will begin. Earth is already the beneficiary of substantial greenhouse effect, thanks to water vapour and trace gases such as methane and carbon dioxide. Certain gases trap heat in the atmosphere because, like greenhouse glass, they are more transparent to incoming short-wave solar radiation than they are to outgoing longer-wave infra-red radiation from the warmed Earth. If there were no greenhouse effect – a point lost on some environmentalists – our planet's surface would have an average temperature of –18 °C, rather than the balmy 13 °C we now enjoy.

A billion years from now we will have too much of a good thing. A vicious cycle will develop, when warming oceans release carbon dioxide and water vapour into the atmosphere, which will lead to further heating, which will release more gases to the atmosphere. Within perhaps a few centuries, Earth's atmosphere will be a mixture of nitrogen, oxygen, carbon dioxide and superheated steam. Some bacteria may survive this catastrophe for a while, but soon even they will perish. Eventually, the oceans will boil dry and Earth's surface will resemble the hellish vistas of Venus, where the temperature never drops below 450 °C, and where molten metal frosts condense on the highest mountain peaks.

The end of life on Earth will in many ways resemble its beginning. A few hardy microbes will eke out a living in a 'warm little pond', as Darwin put it ... although the pond will be scalding hot. Thereafter all will be extinct. No animals, no plants, no restaurants of any kind and not even a slime-mould to liven the monotony. This is a sobering thought; it means that the story of life on our planet is already more than three-quarters done. Should some planetary catastrophe occur, wiping out all higher organisms – a massive asteroid strike, for example – there would not be enough time for them to evolve again.

The Solar System will go on. The expanding Sun that has steril-ized the Earth will bring warmth to the frigid worlds of the outer Solar System. For a while, Mars may resemble the world that many astronomers hope it was in the distant past – warm and wet and conducive to life. The trillions of tonnes of ice buried under its soil will be liberated as trillions of tonnes of liquid water. The ice caps will melt, and once again (or for the first time – the warm-wet hypothesis is far from proven) the great Martian basins and canyons will fill with blue gold. A couple of billion years hence, Mars may be an Eden, any micro-bial life, either native or perhaps the descendants of bugs brought by Earthly astronauts aeons before, flourishing on its newly irrigated plains.

Further out, things could get even more interesting. Take Titan, the moon of Saturn for which the word 'enigmatic' could have been coined. As I write, a large and expensive space probe of the type that sadly NASA doesn't seem to want to build any more is exploring Saturn, its ring system and its moons. The most eagerly anticipated results concern Titan. Bigger than Mercury and Pluto, if this body weren't in orbit around a planet it would itself be considered a planet. It is the only moon in the Solar System with a substantial atmosphere – ten times denser than ours and with a surface pressure 1.5 times that on Earth's surface. Something like this atmosphere, mostly nitrogen with a few hydrocarbons mixed in, is thought to have enveloped the primordial Earth. Sunlight interacts with these petrochemicals, creating a thin, almost opaque orange haze which shields the moon's surface from sight.

At last, our best telescopes and the cameras on-board the *Cassini* spacecraft are getting a peek under Titan's veil. In January 2005, one of the most spectacular events in the history of space exploration took place when the *Huygens* lander – which had piggybacked to Saturn aboard *Cassini* – parachuted through Titan's murky atmosphere and took a series of several hundred astonishing pictures of its hitherto-unseen surface. Large areas of bright upland are covered in what look like inky-black drainage channels, and there are dark plains stretching for hundreds of kilometres. Pictures taken on the surface show small ice boulders on a loose, clay-like surface that may be wet with methane rain. And hundreds of kilometres beneath this exotic surface may be a planet-wide ocean of liquid water. All in all, Titan is a strange and wonderful place, and with its nitrogen atmosphere rich in organics the moon resembles a deep-frozen version (–180 °C) of the early Earth.

But that is now. Four, five billion years hence, with the Sun pumping out maybe a hundred times more energy than it does today, Earth will be long dead, and the outer Solar System could be flourishing. There might be time for life to evolve on Titan (perhaps even the exotic monsters which a few scientists have dared speculate live there even today), for Mars to be rejuvenated, or for Europa's icy depths to be unmasked. The last billion years of the Solar System could see an extraordinary new dawn, a second genesis on the outer worlds, although, if the history of life on Earth is anything to go by, a billion years probably won't be enough for anything more exciting than microbes to develop. It seems that the only chance that intelligent life could ever find a home on Titan would be if our distant descendants – or whatever species has supplanted us – decide to make a run for it before Earth gets too warm for comfort.

Eventually the Sun will swell into a red giant. Ancient, bloated and decadent, it will burn its fuel ever faster. At its peak, our star will be 2000 times brighter than it is now, probably frying anything alive even on Titan. Then the Solar System's true deep freeze, the moons of Uranus and Neptune, plus the double planet Pluto–Charon, may be relatively hospitable oases.

Soon after that the Sun will start to cool, burning helium rather than hydrogen for a hundred million years. Our battered planet, if it survived the inferno intact (which we now think is quite likely) will slowly cool and its surface will re-solidify. The Sun will end its days as an ember, a weakly glowing white dwarf, its retinue of planets deep-frozen for aeons.

That is not the end. It is not even the beginning of the end nor the end of the beginning. To look to the true end of the Universe we must travel a great deal further ahead into time than a piffling five billion years. For on the cosmic scale, five billion years is less the blink of an eye, than the merest twitch in the nerve that precedes it.

One man who has thought a lot about the end of time is Fred Adams. With his colleague Greg Laughlin, Adams wrote one of the most entertaining science books of the late 1990s, *The Five Ages of the Universe*. Adams and Laughlin peer into the far distant future, beyond the last days of our planet into the trillions of years ahead. They conclude that there isn't a great deal to look forward to.

To grasp the enormity of past time is bad enough. But at least the 13.7 billion years thought to have elapsed since the Big Bang is a number that most cosmologists and even some ordinary people can get their heads around. Thirteen billion is the sort of number chancellors talk about. There are people with that much money in dollars and even pounds. It is 13 wheel-barrows-full of sand grains, or five times your age in seconds when you die, if you are lucky. The beginning of time is long, long ago. But it can be comprehended.

The end of time is another matter entirely. So Fred Adams has come up with the concept of the 'cosmological decade': a logarithmic time-scale to cope with the ridiculous numbers associated with eternity. If t is the time in years since the Big Bang, any given point in time hence can be written down as the exponent of the power of ten in years. Thus, 100 or 10^2 years after the Big Bang is the second cosmological decade. One million years is the sixth decade. We are living in the tenth cosmological decade – the Universe is between 10 billion years old and 100 billion years old, which will mark the start of the eleventh cosmological decade. This being a logarithmic scale, time piles up very quickly, yet the numbers remain manage-able. In the eleventh cosmological decade, the Universe will be between 10 and 100 times older than today – i.e. between 100 billion and a trillion years old. That is a vast, unimaginable future, but, as Adams and Laughlin point out, time hasn't even begun to think about getting into its stride by then. So what is in store in this vastness?

A battle. Gravity is going to slug it out with entropy – the tendency of any physical system to become more disorga-nized over time. If gravity wins, everything will be pulled together. If it does not, the Universe will continue to fly apart. Until quite recently, many cosmologists believed that gravity would, in the end, get the upper hand. At the moment, clus-ters of galaxies are zipping away from one another. If there were enough mass in the Universe, the overall gravitational well created by all the matter would eventually reverse this expansion.

So is there enough stuff in the Universe to stop the headlong expansion of galaxies that we see today?

Tricky question. First we need to weigh the Universe and compare the results with what we see – the stars, galaxies, dust clouds and so on. In theory, this is surprisingly easy: astronomers weigh galaxies, say, by observing the effects of their gravitational fields on objects nearby. In practice something odd seems to be going on. When you calculate the velocity of our Sun in the Milky Way and try to account for this knowing the position of all the nearby stars and dust and so on, there isn't nearly enough visible matter to explain its movement. It is a bit like the way Neptune was discovered. Astronomers knew the orbit of Uranus wasn't quite right, that something had to be out there nudging Uranus with its gravitational field. That unseen thing turned out to be a perfectly respectable planet. But when it comes to the stars and galaxies it was mooted as early as the 1930s that something very mysterious seems to be out there.

In short, there must be several times more matter in the cosmos than we can see to account for the observed movements of the stuff we *can* see. This invisible stuff has been dubbed, for obvious reasons, dark matter. Unlike Neptune, its true nature remains a mystery. Most astronomers believe that it is some sort of particle, or family of particles, that only interacts with ordinary matter gravitationally. There could be clouds of these things whizzing through every cell in your body and you would never notice. But it gets worse.

In 2003, the question of whether the Universe would continue to expand forever appeared to have been answered by NASA's Wilkinson Microwave Anisotropy Probe (WMAP). This sits around a million and a half kilometres from the Earth, held in a stable orbit by the balanced gravitational pulls of our planet and the Sun. Launched in 2001, WMAP used a pair of very sensitive telescopes to measure tiny irregularities in the cosmic microwave background radiation, the all-pervasive glow left over from the Big Bang. The probe produced a map of these variations when the cosmos was less than 400,000 years old. The map showed that the Universe was far stranger than we had imagined.

It indicated that the part of the Universe that we can observe contains exactly the right amount of mass to prevent the Universe from collapsing. All other things being equal, it will carry on expanding forever at an ever-decreasing rate. But even if you add together the matter and dark matter, there still isn't enough stuff to make the Universe the shape it appears to be. There has to be something else entirely filling the gap. That something is the strange anti-gravity force 'dark energy'; and it looks to constitute some 70% of the Universe.

To recap, the cosmos is made of three kinds of stuff. There is ordinary or 'baryonic' matter: atoms and electrons and so on, the normal stuff (and its energy equivalent, for as Einstein showed, matter and energy are just two sides of the same coin). This makes up just 5% of the Universe. Then there is dark matter – still stuff, but deeply mysterious, giving another 25%. And then there is dark energy, which is quite as sinister as it sounds. Seventy per cent of everything that exists is this strange pervasive force field.

If the WMAP data were wrong and the Universe, by some mathematical quirk, turned out after all to have enough stuff to rein itself in, what would we see? Compared to the quasi-eternal time-scales of an ever-expanding Universe, a recollapsing 'closed' cosmos exists for just a fleeting instant. If the expansion is to stop it won't happen for at least another twenty billion years. Our Sun by then will be long dead, as will a good proportion of the stars we can see in the night skies. Smaller, slower-burning stars like our nearest neighbour Proxima Centauri will still be aglow. And of course new stars will have formed.

Remember, this probably isn't going to happen, but let's consider anyway the continuing evolution of the closed Universe. In 33 billion years, the Universe is about twice as big as it is today, and the faint radiation from the Big Bang is half its current temperature of about 1.4 K. This era would represent the Universe at its most entropic.

What happens next is interesting. The stuff of the cosmos starts to fall back together. Not individual stars, but slowly and ever-faster the distant galaxies stop receding and start to get closer. The sky starts to draw in. Twenty billion years after the cosmic elastic is stretched to its maximum, the Universe is about the same size as today, with roughly the same density of matter. The large-scale structure of the Universe, the huge filaments and sheets into which the superclusters of superclusters cluster, remains fairly intact for another ten billion years or so. After that the Universe becomes one gigantic supercluster.

If Earth survives to this stage, the skies appear much the same, although anyone with a telescope would notice that something is up. A couple of billion years later, the galaxies themselves merge, as the Universe becomes just one per cent of its current volume. The night sky on Frogstar World B (or any other planet for that matter) is now noticeably brighter, ablaze with stars.

With only 10 million years to go, the cosmic background radiation temperature reaches about 0 °C and there is no longer such a thing as a truly dark sky, save in Krikkit-like planetary systems enveloped by really thick dust clouds, perhaps. Even now, although the views would be spectacular, life is still quite tenable. With the abolition of cold, the swarms of frozen Kuiper-belt-like objects which probably orbit every star would melt. The exotic organic chemicals thought to be locked up in such objects and the countless comets and other icy bodies could spawn new life. It would be a bitter irony if the Universe, relative moments before its death, suddenly blossom on an unprecedented scale.

In our Solar System, or what is left of it by then, the outer planets, the icy moons of Jupiter, Saturn, Pluto, Charon, Triton and Miranda and the rest could become toasty, hospitable places – perhaps for the second time (the first being billions of years earlier when the dying Sun swelled to a red giant). But with six million years to go, most remaining life in the cosmos is extinguished. The night skies blaze brighter than the Earth's day sky today, and the cosmic background temperature exceeds 100 °C – the boiling point of water. Unnumbered ecosystems are snuffed out in an instant, just a few million years after being born.

Now things start to get really spectacular. It is an end like this, by the way, that gives Milliways' diners a brilliant backdrop to their exotic meals and fine wines. (Since life by now is untenable, Milliways' patrons are helpfully protected by industrial force fields and a convenient time bubble that rocks the whole apparatus back and forth across the final moment.) Anyhow: with temperatures rising, the stars start to ignite, their surfaces boiling off into the void. From just about anywhere the view is truly awesome. With only a few hundred thousand years to go, the Universe is in turmoil. Ten thousand times smaller than in the 21st century, the edge just a million light years or so away, the whole cosmos is crammed into a space smaller than that which today separates the Milky Way galaxy from the great Andromeda spiral. The night sky (if there are any solid objects left from which to contemplate a night sky) is hotter and brighter than the surface of today's Sun.

When the background radiation reaches 10 million degrees or so, all the surviving stars more or less simultaneously explode. The entire Universe, at only a few light years across, effectively turns itself into a vast hydrogen bomb. Any solid matter that has survived – the intensely compact cores of gas giants, even the staggeringly robust spheres that are the neutron stars – sublimates into the maelstrom. Atoms themselves evaporate into electrons and protons, as radiation starts to win the battle against matter, just as it did in the few millennia after the Big Bang. With only centuries remaining until The End, even the sub-atomic soup starts to break down. The Universe is pulverized in a gigantic, gravity-fuelled blender.

In the final seconds, Milliways' diners effectively watch a rerun, backwards, of the Big Bang (saving themselves perhaps the cost of a reservation at the Big Bang Burger Bar – see Chapter 6). The forces – the strong, weak, electromagnetic and gravity – separated at birth for tens of billions of years reunite in triumph. Time is, to say the very least, running out. The Universe is now far, far smaller than an atom. In fact, compared to it an atom is significantly bigger than the Universe is today, compared to an atom ... the other way. And, as at the centre of a black hole, the laws of physics start to make their excuses, and leave for a long and well-deserved lunch.

'We don't know what happens next', says Fred Adams. It could be, quite simply, The End. No more matter, no more energy, no more time, no more laws of physics, no yawning void even. The Universe simply gathers up its belongings and exits, stage left. If there is a God, He would be saying, 'Oh well, that's that then', packing away his tools and going off for a quiet sit-down.

Another alternative is that the whole process reverses itself. Maybe a few seconds after everything goes down the pan it all spews forth again. This idea, of an oscillating Universe eternally bouncing between Big Bang and Big Crunch in an endless cycle of birth, death and renewal, is well respected in cosmology (and Buddhism). It is called the Big Bounce.

Enough of this reverie. Remember that for the Universe to collapse we need sufficient matter and energy to generate adequate gravity to overwhelm the expansion of the galaxies that we see today. And look as they might, astronomers simply have not managed to find the necessary quantity of stuff to stop it all flying apart. Not by a factor of 10, or even 100. Even if we dial in all the strange and exotic gubbins we can't see: the dead stars, the black holes, the vast clouds of mysterious dark matter that we know are hiding behind the curtains, and especially the dark energy.

Chilling as the idea of a Big Crunch is, the alternative is in many ways far more terrifying. Also known as the Big Freeze, it supposes an eternity of bleak and extremely depressing noth-ingness. For several trillion years life goes on pretty much as normal. Life, as we know it, and as we don't, is perhaps better and better able to exist. The stars that we see around us are mostly composed of the recycled material left over from super-novae – suns that have exploded over the last 13 billion years. When a massive star dies, the nuclear furnace at its core gob-bles its way up the periodic table, converting hydrogen to helium, helium to lithium, lithium to beryllium, beryllium to boron and so on. It is in the embers of these dying stars that are born the elements we all know and love – carbon, iron, oxygen, sulphur, gold and the rest. We are all starstuff in the most literal way.

There will be much more starstuff in the aeons ahead. In another ten billion years many of the current generation of stars will die, their cores exploding and showering the cosmos with yet more heavy elements. It could be that a golden age of planet formation, when silicon, oxygen and iron are abundant, lies far in the future. What Fred Adams christens the 'Stelliferous Era' will last until well into the 14th cosmological decade – i.e. when the Universe is ten thousand times older than it is today. It may be the golden age of life, too, a true Star Wars universe; we may be living relatively early on in a phase of the Universe's history when something as complicated as life is possible at all. This would explain why we haven't yet found anybody out there. (This is far from certain; rates of star formation may already have peaked, meaning that the pan-galactic civilizations beloved of sci-fi writers lie only in the distant past – another explanation of why we haven't found any aliens: they are all dead.)

By the Universe's hundred trillionth birthday – far, far longer than it would have survived in the Big Crunch scenario – the last stars go out, and the supply of fuel to make new ones is running out. Sipping their hydrogen fuel at incredibly parsimonious rates, shrivelled miserly red dwarfs may eke out a living for trillions of years. If life ever becomes truly common, massive-scale migration will need to be made in every galaxy to these few remaining centres of warmth.

After one hundred trillion years, the Universe is vastly big, and mostly dark. There is, of course, still plenty of stuff. There are all the dead stars, brown and white dwarfs, neutron stars and black holes. There are also planets, gas clouds, dust and an awful lot of that dark matter, the mysterious gloop that physicists know is there but which they cannot get a handle on.

The galaxies may be dark, but they stick together. As the cosmological decades race on, there are new opportunities for excitement in the ageing Universe. Star collisions, for instance, are extremely rare on the sort of billion-year time-scales we are used to. But in the 15th cosmological decade – when the Universe is between a thousand and ten thousand trillion years old – there is enough time to play with for these events to become quite common. Brown dwarfs supply the raw material for passing collisions. Some of these result in no more than two wrecked stars, but if the geometry is right, the two masses of hydrogen merge in an entity big enough for fusion to start. A new sun is born, and in this era any galaxy the size of our own contains about a hundred of these with a combined power output equivalent to that of our Sun. Planetary systems may form from the gas ejected by these collisions, and these may in turn be home to life.

Any beings alive at this time dwell alone in the void. If they develop sufficiently advanced astronomy, they no doubt wonder at the impossibly dramatic spectacle that must have been the Universe in its earliest days – now – when the night skies blazed with light.

From here on, to be honest, it gets a bit gloomy. Slowly the galaxies dissipate, shedding their stellar remnants. Yes, the Universe is occasionally rocked by the collisions between various bits of starstuff. Some of these cataclysms, crashes between white dwarf stars of just the right mass, for instance, produce exotic, short-lived beasts such as carbon- and helium-burning stars. But these events are rare, and as our logarithmic decades race on they get rarer. Eventually even the longest-lived of the small stars formed by brown dwarf collisions run out of fuel and die. Now the only reliable power source of the Universe is the annihilation of dark matter which takes place in the cores of the remaining white dwarfs. For countless trillions of years – twenty cosmological decades – the warmest things in the cosmos are these dark matter burners, whose surfaces radiate at a distinctly unprepossessing –210 °C – considerably colder than the frigid surface of Titan today.

By the 30th cosmological decade, matter itself is in trouble. Protons, an essential component of the atomic nucleus are not, as was once thought, immortal. They last a long time, ten trillion trillion trillion years on average. The physics of the process is still poorly understood, and this number may be on the conservative side, but as Fred Adams puts it, 'in the face of forever, the protons will soon be gone'. And forever is something that an expanding Universe has in buckets.

Proton decay forms the final real fuel for the dying Universe. Like a starving man, a white dwarf remnant that has survived this far consumes its own body in exchange for a few feeble watts of power. As each of the building blocks of matter itself dissolve into radiation, about 400 watts is generated. Thus, in the 37–40th cosmological decades, entire galaxies (or what is left of them) radiate with about the power of a small town on Earth. The Universe is now less energetic than a single star today. Stars which in their youth shone out across the cosmos end their days as a lump of light, transparent solid hydrogen rock about one-seventieth the size of the Moon. The same thing happens, eventually, to neutron stars and any remaining planets. If Earth has survived to this time – which is not altogether impossible – it eventually winks out of existence like the smile on the Cheshire Cat.

Left is a sea of elementary particles and radiation. And black holes. Until quite recently, it was thought that these disreputable monsters (collapsed stars whose gravity is such that escape even for light is impossible) could live forever. The English physicist Stephen Hawking realized in 1974, however, that even black holes are destined to evaporate, just like ordinary matter. As protons will find, true immortality is hard to come by in today's cosmology.

By the 40th cosmological decade black holes are just about all there is. Emitting heat at about one ten-millionth of a degree above absolute zero, they are the hottest game in town. To explain: through the peculiarities of quantum gravitation, the surface of a black hole is not, as it turns out, completely black. A small quantity of the body's mass is always being converted into radiation, and this slowly leaks away into space. When the background radiation falls below a certain level, black holes emit more than they receive (today the cosmic background is significantly greater, so black holes take in more than they give out). As black holes shrink, they get hotter. So in the 40th cosmological decade, for the first time in countless aeons, visible light is again present in the Universe. The dying days of a solar-mass black hole might last for some hundred million trillion trillion trillion years, during which time the city-sized object shines about as brightly as a cinema projector.

By the 67th cosmological decade, black holes have shrunk to a size where they are no longer stable. Blinding flashes of light and heat are seen in the cosmos as black holes explode with more power than today's worldwide nuclear arsenal. The explosions produce clouds of protons, neutrons and the like – by now exotic particles. Matter may once again exist in the Universe, before being nobbled for the second time by proton decay. The biggest black holes may live to the 100th cosmological decade, when the Universe is a googol years old. Some really large ones might actually turn out to be immortal.

The Universe enters its gloomiest phase: the dark era. There is no matter, no heat, no light, no stars or even the remains of stars. Even those doughty survivors, the black holes have (probably) gone. The only objects are the most elementary of particles, the electrons and positrons, neutrinos and photons. And these are spread incredibly thin.

What next? No one knows. There is the depressing possibility that the Universe simply carries on like this for eternity and nothing of interest ever happens again. Alternatively, some physicists think that given enough time – and we have plenty of *that* – a sort of phase transition, like the melting of ice to water, might occur within the vacuum itself, sparking a new universe or universes from the corpse of the old.

There appear then to be two leading possible fates for our Universe – two possible vistas for Milliways' diners to gaze over as they finish their desserts and start to think about coffee and who is going to drive the spaceship home. If our Universe is closed, its end will be showy and will happen in around fifty billion years' time. If, as seems far more likely, we lose the battle against entropy, the *dénouement* will be grim and take a long time coming. But other endings have been mooted. Some physicists believe that dark energy, the mysterious anti-gravity field driving things apart, will strengthen over time. Eventually, they warn, it will rip apart superclusters, galaxies, stars, planets and even atoms in a cataclysmic explosion called the Big Rip. It is quite possible that what we think of as the 'Universe' is just a tiny part of a much grander ensemble, in which case the fate of our corner of Everything might be seen as a merely local affair. Other researchers reject mysterious force fields and ghostly dark matter, pointing to Einstein's dictum that 'physics should be made as simple as possible – but no simpler'.

Come what may, intelligent life will have some job toughing things out. Forget interstellar travel – trailblazing physicist Michio Kaku has concluded that jumping ship will require the creation and manipulation of matter and energies on the galactic scale. Building shields around decaying stars and black holes to conserve their energy, for instance.

One possibility, which Kaku describes in his book *Parallel Worlds*, might be to 'corral neutron stars, which are about the size of Manhattan but weigh more than the Sun, and form a swirling collection of these'. The idea is to create a ring-shaped black hole through which a spacecraft might travel, unscathed, into a neighbouring parallel universe. There would be no way back, but as Kaku points out, 'for an advanced civilization facing the certainty of extinction, a one-way trip might be the only alternative'.

Another option would be to make a new universe and jump in. First you would create 10^{89} of every type of fundamental particle. Then, with fantastically powerful laser beams, you might compress your ingredients to an impossibly small volume. Hopefully a baby universe would form at this point, connected to ours by a temporary wormhole. Anyone jumping into this wormhole before it closes (with a nuclear device-sized explosion) would find themselves in an energetic sea of particles and photons like our early universe moments after the Big Bang. They would then have to sit tight for a few billion years while cosmic evolution took its course and allowed friendly objects like stars and planets to form. At least the view would be more interesting than the bleak void of our dying, native Universe.

Alternatively the development of two-way time travel (see Chapter 9) might enable imperilled civilizations to mess around quite profoundly with the fate of the Universe. Mining matter from the future and bringing it into the present could perhaps even halt the expansion of the cosmos altogether.

Whatever the fate of the Universe and those in it turns out to be, it is clearly worth the price of a meal at Milliways. And once you have seen that, you might as well go to the other end of time to witness the beginning.

6

the big bang burger bar

'I've seen it. It's rubbish', said Zaphod, 'nothing but a gnab gib.'

'A what?'

'Opposite of a big bang.'

The Restaurant at the End of the Universe

Working out exactly what patrons of the Big Bang Burger Bar watch as they down some rather nasty coffee and reconstituted meat, their minds boggling with the conundra of cosmology, raises so many questions. For instance: where did the Whole Sort of General Mish Mash come from? When? What happened before? Or is this a question dismissible (as until quite recently) as a category mistake, akin to asking 'what is north of the North Pole?'. And anyway, aren't accusations of category mistakery just a cheap philosophical trick designed to throw common sense off the scent?

If matter, energy, space and time were created in one colossal explosion, then there was no 'before', just as there is no Platform 9¾ at King's Cross. This, say an increasing number of scientists, is too glib. Despite all it has recently discovered, physics is still inordinately fond of cause and effect. One thing leads to another and so on. When the thing in question is the whole Universe – the majesty of the galaxies, stars, planets and forces that hold it all together – dismissing the cause as a pair of bootlaces is seen as a pretty poor show.

Cue a certain amount of embarrassed throat-clearing. Because for more than half a century the dominant theory as to the origins of everything stated (more or less) that the entire Universe did indeed spring out of nowhere. The Big Bang model has become one of the best known in modern science. Everyone with a brain smaller than a planet has a really hard time getting their head round the realities of relativity. Anyone with a brain smaller than a medium-sized galaxy has real problems with quantum physics. Yet the idea that everything we see burst into existence in a colossal explosion, the likes of which is wholly unimaginable in its stupendousness, has real popular appeal.

Mooted as early as the 1920s, the Big Bang model states that the cosmos began as a point-like ball of energy and matter of almost infinite density which for reasons as yet unclear blew up and has been expanding ever since. At first reading it is a neat hypothesis. It explains, for instance, why the galaxies (actually clusters of galaxies) are rushing away from each other. This was spotted in 1929 by the astronomer Edwin Hubble. He noticed that the more distant a galaxy is, the more the light of its billions of suns is red-shifted, that is, of a longer wavelength than from a similar-sized collection of nearby stars. This red shift was originally explained as a simple analogue of the Doppler effect that makes the wail of an ambulance siren rise and then fall in pitch as the vehicle races towards and then away from you. Now physicists have worked out that the red shift is a manifestation of the growing Universe stretching space itself.

For the Big Bang was not really an explosion which occurred at a point; it was instead a monumental expansion of space, which carried matter along with it. It is meaningless to ask *where* the Big Bang happened: it happened everywhere. When you look into the sky – into the past – you are looking into the fires of the Bang. It is not meaningless, however, to ask *when* it happened; we know that answer, of which more soon.

That the distant galaxies are speeding away from us is not proof in itself that they were once clustered together. For a while an alternative explanation was that new matter and energy are created all the time to fill in the gaps left by the rushing-apart galaxies – the equivalent of a few atoms of hydrogen every cubic kilometre or so per year, say. This was steady state theory, another child of the 1940s, and the Big Bang's great early rival. It assumed that the Universe is the same everywhere for all observers and has ever been so. A steady state Universe is infinite, unchanging, dynamic yet in evolutionary terms essentially static.

Steady state theory was an expansion of the philosophy that dominated the Earth sciences for more than two hundred years until it became clear that catastrophic events such as ice ages and meteorite impacts shaped our planet and its inhabitants. 'We find no vestige of a beginning, no prospect of an end.' So wrote the great Scottish geologist James Hutton in 1785, the man who, perhaps even more than Darwin, sowed the seeds of our modern concept of deep time – the sort needed for the evolution of complex life and mountains. This 'uniformitarianism', whose most famous proponent was another Scottish geologist, Charles Lyell, held that the processes and conditions that we see around us today are the same as those that pertained in the past. It became heresy to posit that past events could be explained in terms of one-off cataclysms; such thinking was resonant of biblical beliefs in floods and plagues.

Steady state theory was championed by the British physicist and writer Fred Hoyle, yet to others it seemed arbitrary and odd. Where were these new hydrogen atoms coming from? How could they just appear from nowhere? Hoyle countered that a stupendous explosion out of nothing was just as nonsensical. He even coined the expression Big Bang on a televised BBC debate in 1949, as a pejorative term to show up the absurdity of the idea.

The clash between the steady state and Big Bang hypotheses was a clash between uniformitarianism and catastrophism, and also just one more manifestation of the oldest philosophical debate of all. Was it ever thus? Aristotle reasoned, not unreasonably, that nothing could come from nothing. Augustine said that to talk of a time before the Universe was meaningless – God created time, sub-atomic particles, the whole menagerie of leptons and quarks, fermions and bosons (although he didn't quite put it like that). Finally, by the 1950s, these questions, for so long addressed by theologians and philosophers, were firmly the preserve of testable physics.

The steady state hypothesis received a series of fatal blows in the mid-20th century. The *coup de grace* was the discovery, in 1963, of the cosmic microwave background. This was predicted by the Russian–American writer and nuclear physicist George Gamow in the late 1940s and confirmed in a survey of radio emissions from the Milky Way done by Arno Penzias and Robert Wilson. This faint afterglow of the Big Bang is a three-degree-kelvin bath of heat around every star and galaxy in the visible Universe.

Another strong plank of evidence for the Big Bang came from observations of a 9:1 ratio of hydrogen to helium in the Universe, a discovery made by various astronomers. Back in the 1940s, Gamow and his colleague Ralph Alpher calculated that no more than three minutes post-Bang, the primordial soup of sub-atomic particles had cooled sufficiently to allow matter to form from stable protons and neutrons. They reckoned that this Big Bang nucleosynthesis produced hydrogen, helium and lithium in the exact abundance ratios that astronomers found. (Other elements – iron, carbon, gold and so on – came later, forged in exploding stars.)

Finally, as telescopes got bigger and better in the mid-20th century and new ways were found to explore the cosmos – using different wavelengths of electromagnetic radiation – it became clear that the Universe a long way away (i.e. in the distant past) is very different from today's Universe. Ten billion years ago it was full of quasars – energy sources that can outshine a galaxy yet are no bigger than a star. These days quasars are rare, perhaps extinct. In other words, the Universe is not changeless and eternal as Fred Hoyle argued. It has evolved.

So the Big Bang is, probably, how it all began. Unfortunately, agreeing on that doesn't tell us where the Universe came from, why or what happened before ... if such a concept has meaning. The Big Bang is what happened *after* the Universe came into existence. As MIT cosmologist Alan Guth elegantly puts it, 'The Big Bang theory leaves out the bang. It tells us nothing about what banged, why it banged, how it banged or, frankly, whether it ever really banged at all'.

So what was the true beginning? In May 2004 physicist Gabriele Veneziano wrote a provocative article in *Scientific American* entitled 'The myth of the beginning of time'. In it he calls for the scientific 'blasphemy' of the pre-Big Bang to be addressed, not ignored. One clue as to what went on before the beginning, he suggests, is what we think happened just afterwards – the blast of colossal expansion called cosmic inflation.

Inflation solves the so-called 'horizon problem'. The basic physical properties of the galaxies are identical wherever we look. Galaxies are made of the same sort of stuff, in the same ratios, as our own Milky Way. The sky is not lumpy but as smooth as a well-puréed soup. Now it is possible that the starting conditions of each region of space in the observable Universe were identical at birth. But this sounds dangerously like coincidence. And scientists don't like a coincidence. An alternative explanation is that the flung-apart galaxies were close enough for their properties to be shared somehow. For this to happen the early Universe must have been much smaller than supposed.

In the past twenty years or so a theory has emerged that 10^{-35} seconds after time-zero the entire cosmos expanded – far faster than the speed of light – until it was at least 10^{50} times bigger, and maybe even infinite. Guth, who proposed the idea in 1979, argued that this inflation must have been driven by some sort of quantum anti-gravity field that he called the 'inflaton'. Immediately following this frenetic explosion, the expansion rate settled down to that we see today. In short, most of the really interesting events in the evolution of the Universe took place in the first, tiniest fraction of a second. Confirmation came in the early 1990s, when pictures taken by NASA's Cosmic Background Explorer Satellite (COBE, the first spacecraft to be dedicated to cosmology) and then WMAP showed the sky as a patchwork of hot and cold spots. These minuscule differences may be the vestiges of minute whirls and eddies created when the cosmos was just a trillion trillion trillionth of a second old, and a similar fraction of a millimetre across – now writ large over billions of light years. The post-Bang inflationary Universe, Veneziano argues, might be a sort of mirror image of what happened before, with an endless series of phase transitions where space–time accelerates and decelerates with no beginning and no end.

The Big Bang also shattered the beautiful symmetry of the primordial singularity, when the four fundamental forces – electromagnetic, weak, strong and gravity – were at one with the fundamental particles. And it destroyed the equality of matter and antimatter. Particles with opposite but identical charges to those we see today should have been created and in equal quantities, physicists calculate. But natural antimatter is nowhere to be found. Where did it all go? In the 1970s Russian physicist and dissident Andrei Sakharov suggested that a small imbalance in the amount of matter and antimatter may have been created in the Big Bang, allowing matter to gobble up the anti-stuff. But we don't really know.

Singularities are uncomfortable entities, infinitely dense concentrations of matter and energy, where the laws of physics break down. At a singularity, for instance, the rules which govern the sub-atomic quantum world must be compatible with those of gravity and space–time (Einstein's parish), something that currently has physicists scratching their heads. One place we can find singularities is in the centres of black holes, and the traditional idea of the Big Bang is as a sort of exploding black hole. This leads to the intriguing possibility that offspring universes are being created within every black hole in our own.

But back to the beginning. It is possible to construct all sorts of pre-Big Bang scenarios, in which the laws of physics and the physical constants themselves evolve and change over time. In one, the Universe has existed forever, essentially as a diffuse void containing distantly spaced elementary particles a lot like the near-eternal near-nothingness in store for our currently all-singing Universe. Then something happens. A transition occurs causing matter to clump together, eventually creating a series of black holes. In these, violent expansions – inflationary episodes of space–time – occur, producing many big bangs of which our Universe is just one result. In this scenario, says Veneziano, 'the pre-bang universe was almost a perfect mirror-image of the post big-bang one. If the universe is eternal into the future, its contents thinning into a meagre gruel, it is also eternal into the past'.

Actually, according to our old friend, Fred end-of-the-Universe Adams, there are subtle differences between the sort of phase transition which Veneziano is talking about and the sort of phase transition on view at Milliways. 'It's more like our universe was born out of a roiling, boiling inferno of a previous high energy space–time', he says.

Another model suggests that our Universe came about as the result of a collision between two precursor universes, aka 'branes', floating within a higher-dimensional space. The model is called the Ekpyrotic Universe from the Greek *ekpyrosis* meaning 'conflagration', and is championed by Neil Turok of Cambridge University and Justin Khoury of Columbia among others. Here two universes with four-dimensional space–time properties collide within a five-dimensional space–time. The result is the Big Bang.

And then some: the bashing branes idea posits a sort of hyperdimensional megaverse in which branes periodically clash and rebound like cymbals. The collisions are Big Bangs, and during the rebound stage the branes expand, accelerating as the turnaround stage approaches. It is possible, this theory says, that the current observed acceleration in the expansion of the Universe is a sign that another collision is on the way. That would shake things up a bit.

Most cosmologists now seem to think that the birth of the Universe was far more complicated than a simple explosion that took place 13.7 billion years ago. According to Leonard Susskind of Stanford University, the initial inflationary phase can be thought of as a sort of 'super bang' which created a cosmos almost infinite in extent. The 'Big Bang' that gave rise to our corner of existence was a local affair – 'one of the bubbles in a bottle of champagne'.

The Big Bang, at least in its original incarnation, is graspable by most people once they sit down with a cup of tea. The same cannot be said for much of what physics has come up with since. Superstring theory and its turbocharged sister M-theory, for example, with their talk of '11 dimensions of space–time' and 'Calabi–Yau shapes' are incomprehensible to anyone without a profound understanding of mathematics.

Since its Enlightenment beginnings, modern science has been within the ken of large numbers of intelligent people, at least in some simplified form. There is nothing in Newton's equations to trouble a bright high school student. The greatest scientific treatise of the 19th century, Darwin's *The Origin of Species*, was written for the public as well as fellow academics. Even relativity, a byword for eggheadery, boils down to a few straightforward equations describing simple quantities like velocity, mass and the speed of light.

It is no longer so. In their quest for a Grand Unified Theory to integrate the quantum and relativistic worlds, researchers are resorting to models, analogies and equations so complex they even befuddle experts in the lab next door. What, in the concrete as opposed to the mathematical sense, is a brane, for instance? What does it mean to talk of our Universe 'floating in higher-dimensional space'?

One of the most successful at bringing string theory to the masses is Brian Greene. What Greene describes in his blockbuster book *The Elegant Universe* is a strange, almost metaphysical place, where whole dimensions can hide behind a single atom, and where space and time may turn out to be illusions, the existence of these fundamental properties being merely the result of some tweaking in the cosmic harmony. A place where the Planck length (10^{-35} metres), the smallest bit of anything that there can be, may conceal the shards of fundamentalness that give rise to everything we know. Greene writes like a dream, but still, most of us need a different brain to have any hope of understanding all this – plus a new language.

It seems that natural selection and relativity may have been the last profound theories that could be grasped by the person in the street. Which is a great shame. It leaves us, as the comedian Ken Campbell cheekily points out, having to 'take science on faith'.

That's the bad news: cosmology involves some incomprehensible ideas. The good news is: it's no longer all theories and no data. Spacecraft like the Hubble space telescope, the new Keck telescopes in Hawaii, and the satellites COBE and WMAP, have captured unprecedented pictures of the very early Universe. Soon particle physicists will probe the earliest moments of the Big Bang itself in the laboratory by creating fantastically high energies. To do this researchers smash particles together in giant accelerators such as the 27 kilometre CERN ring which runs under the Franco–Swiss border near Geneva. To date, accelerators have managed to replicate the conditions that pertained when the Universe was just one ten-millionth of a second old. This is positively geriatric compared to the eras most cosmologists hope to espy with new and existing colliders in the coming decades.

More than half a millennium ago Copernicus toppled humanity from its pedestal. By realizing that the Earth revolved around the Sun as opposed to the other way round (and furthermore did so in the company of other planets, like Venus and Mars) he made our world just like one any other. After astronomers realized that our Sun was a star like any other, our place became even more insignificant. The revelation that there are innumerable galaxies just like ours shrunk our significance again. Darwin relegated *Homo sapiens* to being just another species, and the great founding fathers of geology, men like Hutton and Lyell, consigned our very age to an unremarkable instant in an almost unimaginable calendar of deep time.

When Adams wrote *Hitchhiker's* it was already clear just how insignificant we were:

Far out in the uncharted backwaters of the unfashionable end of the Western Spiral arm of the Galaxy lies a small unregarded yellow sun.

Orbiting this at a distance of roughly ninety-two million miles is an utterly insignificant little blue green planet whose ape-descended life forms are so amazingly primitive that they still think digital wristwatches are a pretty neat idea.

Now science may have claimed its most spectacular scalp with the realization that our Big Bang (and the entire Universe it gave rise to) may have been just one of those things.

7

time travel

There is no problem involved in becoming your own father or mother that a broadminded and well-adjusted family can't cope with ... it all sorts itself out in the end.

The main problem is quite simply one of grammar ...

The Restaurant at the End of the Universe

Time travel, in the *Hitchhiker's Guide*, has increasingly become regarded as a menace (or will have increasingly become to have been seen as a menace, or had become to will have been regarded as a menace – you see the grammar problem?). History is being polluted and we are told a cautionary yarn that anyone even thinking of trying to mess around with tame black holes and wormholes of negative energy should read and digest thoroughly, preferably before breakfast.

This sad tale concerns a poet, Lallafa, who wrote what are widely regarded throughout the Galaxy as the finest poems in existence, the *Songs of the Long Land*. Lallafa lived on a remote, pre-industrial planet, in the Forests of the Long Lands of Effa, far from the centre of galactic civilization. He lived there and he wrote his poems there, on dried Habra leaves without the benefit of word-processing software, laser printers or correcting fluid.

Several centuries later, shortly after the invention of time travel, some major correcting-fluid manufacturers wondered whether his poems might have been better still if he had had access to some, and whether he might be persuaded to say a few words to that effect. Lallafa was by then, of course, long-since dead. Using one of the new-fangled time machines, the correcting-fluid marketing men took a trip back and found Lallafa. They did some tortuous and eyebrow-raising explaining, waved an awful lot of what they managed to persuade him was money and eventually convinced Lallafa to endorse their product. Unfortunately, they found him just as he was about to embark on his career as a poet – i.e. before he had put pen to paper. Thanks to the correcting-fluid-makers' *largesse*, Lallafa became rich and famous and never actually got round to writing any poems. For a start, the girl from whom his love was so famously unrequited took one look at his bulging wallet and took the plunge, thus removing one major source of poetic inspiration. No matter, the correcting-fluid manufacturers thought. They simply packed him off with a later edition of his poems and some dried Habra leaves on which to copy them (and some bottles of correcting fluid of course). Thus the poems were written, and the circle was squared.

But, of course, it wasn't. The story of Lallafa neatly illustrates one of the most worrisome paradoxes of time travel. For while going back in time and shooting your grandmother (or actually becoming your grandmother) is bad enough, it isn't, perhaps, as nasty, in logical and philosophical terms, as the creation of a *djinn* – an object or idea created by the action of time travelling itself. The paradox is this: where, exactly, did the poems come from? They were thought up by Lallafa, so at first the answer seems obvious: from inside his head. But time travel throws a monumental spanner in the works. In the story, Lallafa never actually got a chance to write his poems. Follow the story round, and round, and it is clear that time travel has produced a horrible loop, a temporal roundabout into which the poems just barge, like a London bus, taking no account of the traffic already on the gyratory.

Take another *djinn*. You build a time machine and jump in, having programmed the thing to take you five days into the past. You jump out and find a passer-by, and hand him a pound coin from your pocket. You tell him about the time-travel shenanigans and instruct him to bring himself and the pound coin to a certain place and a certain time – the place, and time, of course where you started out on your first time-journey, five days hence. He duly follows the instructions, and meets you and – of course – hands you the pound coin. This is the same pound coin which you slip into your pocket, and travel back in time with to give him. From where did the coin come? Thin air, it seems. By travelling in time you have created a loop of illogicality which can not only make you your own father (or mother) but can also conjure up objects (the pound coin) and, perhaps even nastier, information (Lallafa's poems) out of nothing. This is a free lunch, and physics abhors free lunches even more than vacuums.

Actually, physics does these days allow for objects popping into the Universe out of nowhere – the virtual particles that are thought to infest the vacuum soup – but these are polite enough to disappear after a fraction of a nanosecond. Hanging around long enough to wear a hole in your pocket and possibly boost the inflation rate simply will not do ... imagine if it had been not a pound but a billion pounds.

Information is worse. Say you showed Einstein the special theory of relativity just as he was settling into his coffee break at the Berne Patent Office back in, say, 1903, two years before he actually came up with the thing. He might try to run from the room with his hands over his ears shouting 'la-la-la' (being Einstein, he would probably have been quicker than most to appreciate the logical obscenity about to unfold). But you could have grabbed him and shouted his conclusions, line by line, down his earhole.

Like energy, information cannot pop out of nowhere. Nor, it seems, can it simply disappear. In 2004 Stephen Hawking retracted his 1970s conclusion that when matter falls into a black hole it is destroyed along with all the information that goes with it – its spin, charge and so on. Black holes are leaky, and information manages to escape their grasp at the last minute, apparently. The introduction of information in the Universe from left field goes so horrendously against the grain of everything that we know to be right and sensible that this phenomenon could quite rightly, as physicist David Deutsch has pointed out, be considered a miracle. Another thing physicists don't like.

The paradoxes of time travel are well acknowledged by science fiction writers, but few have been as willing as Douglas Adams to acknowledge them, shrug their shoulders and carry on anyway. By contrast, in the film *Back to the Future* the becoming-your-own ancestor (in fact, becoming your own father in this case) paradox is addressed quite brilliantly. In the movie, young Marty McFly is transported into the 1950s and ends up becoming embroiled in the love life of his parents. Specifically, he has to make sure that his teenage mother, who has inconveniently and embarrassingly fallen for him, gets it together with his father before the latter is beaten to the finishing post by the smalltown thug who is making everyone's lives a misery. If the thug gets the girl, Marty is history – or rather, isn't history. As Marty meddles with his parents' hormones, his existence becomes less likely. The director has a neat way to illustrate this: Marty himself starts to lose physical definition. He feels ill and at one point he even disappears from a photograph. The present rewrites itself in a desperate attempt to keep up with history, with vigour of which Joseph Stalin would have been proud. This is one solution to the key paradox of time travel – if you change the past, the future simply changes to keep up. But when you think about it this is no solution at all ... if Marty McFly airbrushes himself out of history, as it were, then who is there to do the airbrushing?

In Robert Heinlein's short story *All You Zombies* our hero turns out not only to have been his own father, but his mother too – and the person to whom he is relating his story. When time travel meets gender realignment surgery, anything is possible. And in the *Terminator* movies, the information paradox forms the basis of the plot. In the near future, intelligent machines trigger a nuclear war which almost wipes out humanity. The survivors, led by John Connor, are besieged by belligerent, intelligent robots intent on finishing the job. One of their most cunning plans involves sending a robot, a terminator, back in time to finish Connor off before the war, and before he grows up to lead the resistance. Over the course of the three movies various attempts are made to do this, all without success. But we also learn, in the second film, that the technology to build the new breed of super robots came, originally, from a piece of the first terminator sent back in time. This is a classic example of an information paradox. The technology – the information – came from nowhere.

For decades physicists have been arguing about whether time travel is allowed, and if so, how. (Many would argue that time travel into the past, at least as far as the 1970s, can be readily achieved by walking into any British high street Post Office, but that is another matter). Some physicists get quite snooty about time machines, referring to them instead as 'closed, time-like curves' (CTCs). But despite this snobbishness – which is fast disappearing as the discipline makes its bid for sexiness – it seems at least probable that time travel may not be banned by the laws of physics. All you need do is find a way round the speed of light.

That is the good news. The bad news is that backwards time travel, if possible, will almost certainly be Very Hard Indeed. It is one of those engineering challenges that theoretical physicists wheel out every now and then to show off to their more practically minded friends mucking around in the lab. These challenges tend to involve instructions like: 'First take Jupiter and crush it down so it fits in a box small enough to squeeze under the bed ...' or 'find a black hole with a mass of a million suns and fire a series of neutron stars at it ...'. (That said, there is one man, of whom more later, who thinks that time travel might actually turn out to be quite simple.)

We are talking about backwards time travel here. Forwards jaunts are, of course, trivially easy. We do it all the time, at the trite rate of one second per second (unless you are bored, in which case it is much slower). It is simple to increase this rate. Time dilation might be irreversible but it is well-understood, wholly testable and absolutely works. Einstein showed that no two things moving relative to one another can occupy the same time frame. A fast-moving object experiences time at a different rate from one travelling more slowly or not at all. Special relativity says that the combined speed of motion through space plus motion through time of any object is always equal to the speed of light. The faster you travel through space, the slower you travel through time. The whole idea is summarized in the phrase 'moving clocks run slow'.

If you synchronize watches with a friend on the tarmac, climb into a military jet, blast off at Mach 2 for an hour and then land, you will find your watches are out of synch – yours will be some millionths of a second behind your friend's. You have travelled forward in time, albeit on a one-way trip. Hyper-accurate atomic clocks aboard spacecraft and supersonic planes have confirmed the effect. The people who have experienced the greatest time dilation were the cosmonauts aboard the Mir space station; some of these hardy souls spent more than a year whizzing around the Earth at more than 27,000 kph. On a long mission they gained nearly four seconds on their Earthbound colleagues – a difference detectable even with a cheap quartz watch. In fact, even walking across the room puts you in a different time frame from someone sitting on the other side. Time dilation illustrates the odd notion that there is no such thing as absolute time – we all operate inside our own temporal universe.

A few seconds are one thing. But say you want to travel to AD 2,000,000. You don't need a tame black hole, or a suitcase-sized Jupiter, just a nippy spacecraft. One that can go faster, say, than 99.9999999% the speed of light. As you approach 1000 million kph or so, things start to get interesting. The plot of the time dilation equation is an asymptotic curve – a little like a hockey stick. At low speeds, the effect is marginal. Even at 80% of c, clocks run at about only 70% speed. You really have to go very close to the speed of light indeed before the effect becomes pronounced. At 90% of the speed of light, time for you moves twice as slowly as for a stationary observer. At 99% it runs seven times slower. At this speed, you could get to Proxima Centauri and back – a round trip of eight light years – in a little under 14 months as far as on-board time is concerned. At 99.9% of c your clocks move an hour for every 22 days back on Earth (or wherever you set out from). Add another three decimal places' worth of nines and every hour you experience on your speedy craft equates to two years back home.

Continue to accelerate to $0.99999999999999c$ and for every day on board, nearly 20,000 years pass. This is clearly becoming an effective means of time travel. When your velocity is just one trillionth-part short of c, then every day on board is more than 60,000 years. AD 1,000,000 is less than 17 years away. Go even faster and soon whole millennia are passing for every second in flight. We cannot exceed the speed of light, but within a whisker of it the time dilation effect tends towards infinity; billions of years flash by in an instant.

Time dilation is the saviour of the long-distance traveller. Space is extremely big. To cross our galaxy at near light speed would take a clearly inconvenient 100,000 years. But those 100,000 years are only as measured from 'rest', i.e. Earth.

Shoot off to the Milky Way and back, and as far as your fellow Earthlings are concerned 200,000 years (plus whatever time you spend poking around out there) will have elapsed. But travelling snuggled up to *c*, you could easily make the journey there and back in a lifetime. No need to contemplate hibernation or generational starships, warp drives or hyperspace or wormholes. You wouldn't even need to worry about unfeasibly large acceleration. One *g* for a year or so is quite enough to get almost to light speed, and the same again in the opposite direction to slow yourself down.

Some people ask: how can the time dilation effect occur given that we move in relation to everything else, not to some fixed, universal grid? Surely if Alice zips away from Bob at millions of kilometres an hour, as far as she is concerned Bob is heading away from *her* at the same velocity. So who experiences time dilation? The answer is Alice. She is the only one to have experienced acceleration. Acceleration – not velocity *per se* – is the key.

You might need to worry about whether or not you remembered to turn the gas off, though, because of course your elapsed time will bear no relation to the time back home. It is all very well kissing your spouse and whizzing off to explore the Andromeda galaxy, but it is less good to return to a world where your family has been wormfood for two million years and quite possibly your entire civilization and even species as well.

You can also use gravity as a one-way time machine – it too slows time. Just as every movement puts you in your own time zone, every gravitational interaction an object has with another shifts its time frame a smidge. And if the gravitational field is extremely large, say near to a black hole, then the time-warping effect is large too. On the Earth's surface clocks lose about one microsecond every 300 years. On the surface of a neutron star time runs at maybe half speed. If you sat on one, you would see the rest of the cosmos going about its business in double-time, like in a silent film. Take a pew on the event horizon of a black hole and time effectively freezes: where gravity becomes infinite, time slows down infinitely.

Falling into a black hole would be nasty and quickly fatal as your body would be stretched into spaghetti by the immense tidal forces. There are compensations. Your life would flash before your eyes as doom approached, but everyone else's would too. If you were to fall into a black hole with a mass of several million suns, say, you would pass through the event horizon unscathed; the tidal forces do not get excessively large until near the singularity at the centre. At the point of no return the time dilation effect would approach infinity. From this vantage point you would see stars being born and dying, galaxies evolving and dancing their celestial dance. It would be a magnificent sight – the future in a nutshell. Then you'd disappear from our Universe, and from our time too. It's a 'one-way trip to beyond the end of time', as Paul Davies says so neatly in his wonderful book *How to Build a Time Machine*.

While we're on the subject: it may be possible to delay spaghettification for a short while when falling into even a medium-sized black hole. Deborah Fredman of Harvard University has calculated that all you need is a 12 quadrillion tonne lifebelt. This would create a gravitational field strong enough to balance the tidal forces on your body as you plummet through the event horizon. The ring would buy you just under a tenth of a second more to enjoy the view before you and it are torn to atoms.

Travelling into the future would of course be extremely expensive, either via a black hole or at near light speed. As you approach *c*, you need more and more energy to accelerate for the same increase in velocity. To break the light barrier itself requires an infinite amount of energy. All this tiresomeness comes about because when you travel fast you get more massive; as your mass tends towards infinity, so does the amount of energy you need to accelerate further. Relativistic travel is not a good way to lose weight.

Worse, it is a strictly one-way ticket. High speed or extreme gravity might show you the glittering civilizations of Earth two million years hence, but they won't get you back to tell your mates.

A backwards time machine may be possible. On paper. To understand why, we need to get a better grip on time. Most definitions are alarmingly self-referential and our traditional Newtonian notion of 'an absolute now' is clearly misguided. In Einstein's special and general theories of relativity, three-dimensional space is combined with time to form a four-dimensional entity called space–time.

The pioneer of armchair time travel, H. G. Wells, was on to something when his unnamed Time Traveller in *The Time Machine* explained to his guests that

> any real body must have extension in *four* directions: it must have Length, Breadth, Thickness, and—Duration. ... There is, however, a tendency to draw an unreal distinction between the former three dimensions and the latter, because it happens that our consciousness moves intermittently in one direction along the latter from the beginning to the end of our lives'.

Einsteinian space–time consists of 'spatiotemporal points'. Say you are about 1.8 metres high, 45 centimetres or so wide and 25 centimetres thick. Those are your dimensions in space. You also have a temporal dimension of 78 years, say. Thus your life forms a sort of elongated worm, you-shaped in cross-section and tapering at one end (the time you were growing from embryo to adult). At any one time 'you' are a three-dimensional cross-section of that worm.

The line any object follows through time is called its 'worldline', and time, as measured by any device, will always increase in one direction along this line. Worldlines need not be straight; acceleration alters the angle of the worldline with respect to the time axis, as does gravity. Extreme gravity can distort space–time to the point where worldlines can form a closed loop – a corridor to the past. A closed time-like curve, or CTC, in other words.

In 1948, the logician Kurt Gödel calculated that a rotating universe would generate its own CTC and thus allow you to travel into the past. Current observation shows that our Universe does not spin; when you add up all the spin of the observed galaxies they cancel each other out. A CTC might also be produced naturally by a rotating black hole, though as we've seen, black hole travel is rarely worth the risk.

To date the most practical devices dreamt up by physicists for backwards time travel have involved spinning cylinders and wormholes. In 1974, Frank Tipler, Professor of Mathematical Physics at Tulane University in New Orleans, calculated that a cylinder of incredibly dense matter – say the stuff of neutron stars – set spinning at sufficient speed would effectively wrap space–time in loops with its gravity. It's the same principle as in a Gödel rotating universe. In fact, the gravity of any extremely massive rotating object will create a whirlpool in which any light ray will be whipped round and round back on itself. If you, the time traveller, fly through this maelstrom at sufficient velocity, you can effectively outpace the speed of light relative to an outside observer. Tipler worked out that a cylinder 100 kilometres long and 65 kilometres across should do the trick.

The most physically correct solution to the time travel conundrum to date – the one many physicists are convinced would work if only we had the wherewithal to build the wretched thing – is the wormhole. First proposed in the late 1980s by CalTech physicist Kip Thorne, the wormhole machine obeys Einstein's laws, relies on no magic, and requires you to believe in not a single impossible thing. Thorne's gizmo leaves all the paradox questions unanswered – going back and becoming your own grandfather and so on – and it would be pricey and tricky to build, but it would probably work. Possibly.

In the early 1980s, Carl Sagan asked Thorne to find a way in which an astronaut could be transported to a distant part of the Universe and back to Earth in a matter of hours or minutes. Sagan was writing a novel, *Contact*; its plot revolved around getting a group of people from Earth to the star system Vega, several light years away, quickly. Sagan was keen to obey the laws of physics, and so dismissed the usual sci-fi copouts of 'warp drives' and so on.

Thorne decided that a wormhole is the way. First mooted in the 1920s by Einstein and his student and long-time collaborator Nathan Rosen, wormholes are tubes of warped space–time that bridge two distant regions of space (and, as it turns out, time). A black hole is a particularly tricky type of wormhole, but Sagan needed something much more stable. Something that would remain open long enough for someone to jump through and survive.

To digress for a moment. There *are* ways of messing around with black holes that seem to allow them to be hijacked for backwards time travel as well as forwards. As well as using a large mass to balance the tidal forces you could also add a colossal amount of electric charge, for example, or set one spinning so that the tidal forces on the way to the singularity at the centre are not too destructive. The singularity would then be manifested as a doughnut-like chute through which you could zip unscathed. Unfortunately, allowing travel into a singularity and out the other 'side' is seen as nothing less than pure and undiluted evil by most physicists. Singularities, the crushed point-like nightmare of matter and energy and who-knows-what at the centre of a black hole, are places where left is right, effect precedes cause and no one knows where they are. That is why many people believe that singularities can never be naked and must always be cloaked, for decency's sake, by the impenetrable boundary of a black hole's event horizon.

Far more benign is a wormhole that has formed naturally. These rents in the fabric of space–time may be all over the place. The vacuum, it seems, is far from the nothingness one would imagine. It is a seething mass of energy that the American physicist John Wheeler (who coined the term 'black hole') has christened 'space–time foam'. The average of all this energy is zero – hence nothing appears on the macro scale. But on the minutest Planck scale, virtual particles and huge energies are constantly popping into and out of existence. Some of these will be powerful enough to warp space itself into tiny, fleeting wormholes. The trick would be to get hold of one of these and, by pumping in a gigantic amount of energy, expand it into something large and stable enough to be useful. Unfortunately, the biggest energies we can manufacture on Earth – those created in particle accelerators – are just too low.

Instead, astrophysicist Paul Davies suggests directing the explosive force of a ring of hydrogen bombs at a small ball of quark–gluon plasma – a high-density state of matter consisting of dissociated quarks and the particles that bind them together. This stuff, which can be created by the most powerful particle accelerators, is thought to resemble the chaos that followed the Big Bang. These shenanigans might be enough to create a little something with the sort of density needed to rip a permanent wormhole in space. A density, in other words, that would make neutron starstuff look like candyfloss.

We now have a stable but small wormhole. To inflate it to human-size you need to pump in an amount of energy equivalent to the mass of Jupiter. Where to get this from? Davies suggests that an extremely high-powered laser in a particular configuration could produce enough energy (via a phenomenon called 'squeezed light') if turned on for several trillions of years. Another source of energy would be a black hole, but we want to avoid setting up our time machine factory anywhere near one of these, for obvious reasons. Or the wormhole itself might have a powerful enough gravitational field to generate the required quantities of exotic matter very quickly. In other words, we set off the nukes and watch our ticket to the past grow before our very eyes. It sounds too good to be true, but then so does going back in time and buying today's winning lottery ticket.

Lastly, the wormhole needs to be turned into a time machine. It is no use having a tunnel through space six feet long that connects you with, er, right now six feet away. It would be a waste of time and money – think of all those nuclear bombs and the cleaning up afterwards. Fortunately, this is where we can use a much simpler form of time travel to prime our machine. Grab one end of the wormhole using an electromagnetic field, and put the other end in a giant particle accelerator and whizz it around for a few years at near-light velocity. Thanks to time dilation, the stationary end slowly moves into a different time zone from the moving end. Whirl one end for say a decade and you will get a wormhole to 10 years in the past. Leap in, emerge ten years ago. Neat.

Ronald Mallett, a professor of theoretical physics at Connecticut University, likes to keep things a bit simpler. He thinks one could build a time machine in the lab. With a light beam.

In 2000, Mallett published a paper showing how a circulating beam of laser light could make a vortex of space within its circle. He then had what he admits was a eureka moment. 'I realized that time, as well as space, might be twisted by circulating light beams.'

Like so much else, did H. G. Wells get it right when he described his Time Traveller mounting a machine equipped with a large spinning disc at the rear, through which the fourth dimension could be navigated? Mallet is a big fan of the late Mr Wells. After his father died at the age of just 33, the 10-year-old physicist-to-be was devastated. 'After reading Wells's *The Time Machine*, I became obsessed with the notion of building such a device to see my father again', Mallet recalls. 'That eventually led me to physics and my current research.'

Mallet originally thought that to twist time into a loop he would need a second light beam, circulating in the opposite direction. But his latest research hints that a single one-way cylinder of light (rather like the swirling vortex in the 1960s TV series *The Time Tunnel*) would do the job. If the light increases in power, time and space swap round. Inside the beam, space becomes the one-way street that time is in the ordinary world. Time, on the other hand, takes on the properties of an ordinary dimension in space. If you entered the tunnel (which happily would not involve being diced, as in entering a black hole) you could, in theory, stroll back and forth through time. Admittedly, you would need a laser more powerful than anything humankind has built so far. Still, Mallett reckons his setup to be within the scope of present-day technology.

Maybe Ronald Mallett will become a real Connecticut Yankee at the court of King Arthur. Or maybe not. Maybe the Universe is littered with ready-made time machines. Cosmic strings, not to be confused with the strings of string theory, are postulated strands of almost infinitely dense, space–time-warping Big Bang debris, scattered through space. Fly round and round a pair of these and each circuit would take you a little further into the past. Better still, distort space into a fold, says theoretician Miguel Alcubierre, of the University of Wales. Like *Star Trek's* warp drive, such a slot would offer faster-than-light passage between two points. It would of course double as a time machine.

So what about those paradoxes? In a *Scientific American* article written in 1994, the philosopher-physicists David Deutsch and Michael Lockwood listed various ways out.

One solution is that all the new causal loops introduced by time travel are logically consistent. You can go back and become your grandfather simply because you *always were* your own grandfather (check those eyes in the mirror – far too close together). If you go back and kill your mother before you were born it will simply turn out that you were adopted. But what happens when, armed with this knowledge, you decide to rebel against history? Suppose you go back to meet your earlier self. At this meeting your younger self records what was said at the meeting and, in due course, having become that older self, tries deliberately to do something different. 'Must we suppose, absurdly, that she is gripped by an irresistible compulsion to utter the original words, contrary to her prior intentions to do otherwise?', ask Deutsch and Lockwood.

Thr common sense answer is yes. Something must go wrong to prevent this happening. The time machine will break down, or our time traveller will have a cerebral seizure forcing her to utter the original words. As Paul Davies explains, 'unfettered free will' is at the heart of the problem. If, as most rational thinkers now accept (because not to do so seems to invoke fairies at the bottom of the garden), free will is an illusion, this could remove some of the objections to time travel: a non-contingent decision to kill your own grandfather is simply not possible. To be on the safe side, Stephen Hawking came up with the 'Chronology Protection Conjecture' that defends the cosmos against the pollution of history. In a nutshell the conjecture states that time travel, while theoretically possible, will be practically impossible because the Universe will protect itself from time machines by not allowing them to be made.

We do not live in a classical, common-sense Universe. The Newtonian dynamics that works so well on the scales we can appreciate is only an approximation. The real world, where quantum minutiae jostle relativistic might, is much more messy.

There is one good thing, however, about quantum physics: it offers a neat way out of the time travel paradoxes. If an electron, say, can 'choose' to go left or right after a collision, with not a thought in the world about the state of affairs just before, then perhaps at that point two quite independent worlds, or universes, are created – one where the electron turned right, the other where it turned left. The many worlds hypothesis was first postulated by Hugh Everett III (who studied with John 'black hole' Wheeler in the 1960s). It states nothing less than that we inhabit one of an infinite number of parallel universes – an idea which crops up time and again in the *Guide* and deserves a chapter of its own (Chapter 12).

Imagine you go back in time and successfully assassinate Hitler in 1933. There is no Nazi nightmare, no Second World War. Millions of people who would otherwise have died are saved. Many other people, though by no means as many, hopefully, would die instead of living (Hitler, for a start). Clearly this would be a momentous event that would irrevocably change the course of the future, including possibly the circumstances that would have led to your conception and birth. You travel back to the future and find a different world, one perhaps under Soviet domination, or American military dictatorship, or a peaceful pan-global democracy. Maybe this world contains no trace of your ever having existed. It doesn't matter. You do not suddenly disappear, nor is there a problem when you return to the future.

By travelling back in time and altering the past in some way you merely alter the past of a new, parallel world. There is no paradox and no inconsistency, because in this universe, the one in which your Hitler is assassinated by you, you had never existed and never will exist. You are an alien interloper from a parallel world. *Djinns*? No longer paradoxical. The pound coin that seems to come from nowhere was minted in one universe and taken to another. If you go back and tell Einstein all about his *annus mirabilis*, 1905, it doesn't matter either. In *this* universe, that was how Einstein found out about it. In *your* universe, Einstein's ideas were his own. You have mined information from one universe and given it to another.

Alternatively there's what we will call, for argument's sake, the Yoga solution. The British physicist Julian Barbour, in his thought-provoking book *The End of Time*, suggests that our notion of past, present and future is merely a property of the way our brains have evolved and grown. The customary view of time is as a river, down which we are all propelled by an unstoppable and irreversible current. He recommends envisioning it as a sort of Platonic solid, in which events are embedded. In this model, time is 'nature's way of stopping everything happening at once'. The present, future and past can all be said to have coincided. A comforting thought: if the aeons before our births coexist both with our lives and also the countless aeons afterwards, then death ceases to have meaning. It is simply a part of the Whole Sort of General Mish Mash.

'To tell the truth, I find the idea of time travel boring compared with the reality of our normal existence', writes Barbour. 'Our memories make us present in what we call the past, and our anticipations give us a foretaste of what we call the future. Why do we need time machines if our very existence is a kind of being present everywhere in what can be? We are all part of one another, and we are each just the totality of things seen from our own viewpoint.' Which is all very well, but not much help if you want to go and watch the Battle of Hastings.

There is circumstantial evidence that some of the postulated time machines may never be built, at least by our descendants. We can assume that we live in quite an interesting period of history. This is not just temporal vanity or chauvinism. Plenty of other periods can lay claim to being interesting. Parts of the Middle East were worth looking in on 2000 years ago, for instance, when there was the golden age of the pharaohs, the Romans, Classical Athens and so on. But the 20th and 21st centuries will surely always be seen as special, for all sorts of reasons. It was the era when humans learned to fly in space. We invented computers and, for the first time, warfare became something that could threaten our entire species and even our planet. Most importantly, the late 20th century saw the dawn of the world's first true global civilization. Not under one leader, of course, or able to agree about anything, but there were the beginnings of a shared culture that cuts across even the fault lines of war and territorial aggression. The Internet, international commerce and cheap, fast air travel mean that the world is a smaller place than it has been before. All of this together, I would argue, adds up to a pretty convincing reason for why anyone in the near or distant future would want to visit our times and have a look.

So, where are they all? Where are all these tourists from the future? And why don't we have reports of inappropriately dressed visitors at other key events in the past? One answer is, as some designs for a plausible time machine suggest, that even if time travel proves possible, it will never be so to a point further back in time than when the first time machine was built. This is the case with wormhole time machines, which are anchored firmly at the time they are constructed. Another explanation is that time travel is possible, but simply that we humans never get round to it. Maybe we can't be bothered, or we aren't clever enough to do the maths, or maybe our civilization collapses before we get a chance. Other alien civilizations may develop time machines, but they, presumably, would be mostly interested in visiting their past, not ours. Or that if the many-worlds hypothesis holds true, our Universe is simply one of the trillions not to have been visited by time travellers from the future. If time travel does turn out to be easy and common-place, we may well find that when we finally strike out and explore our Galaxy the whole thing has been through the temporal meat grinder. In which case, we may wish to join the Campaign for Real Time.

There remains the possibility that humans will develop time travel in the not too distant future and furthermore use their time machines to travel arbitrarily large distances into the past. Then the answer to the question 'Where are all the time tourists?' might simply be 'They are here'. I have never for a minute believed that the Earth is being buzzed by alien life forms in flying saucers. There seems to be no convincing argument that alien life is not quite common across our Galaxy, but also no convincing reason why they should choose to visit us. But our descendants, in their time machines, might. It is only a small chance, but it is just possible that among all those fakers and charlatans are a few people who may just have caught a glimpse of our future.

8
the babel fish

The Babel fish is small, yellow and leech-like, and probably the oddest thing in the Universe ... if you stick a Babel fish in your ear you can instantly understand anything said to you in any form of language.

The Hitchhiker's Guide to the Galaxy

To anyone who has struggled, and failed, to learn a foreign language, a real Babel fish would be a boon almost beyond compare. Imagine all those holidays in France where you could order a rare steak and actually get some slightly charred cow, instead of the boiled otter which the waiter insists you asked for. Picture going to the cinema and smugly watching the latest Kung Fu fest without having to crane over people with big hair to glimpse the subtitles. Imagine being able to understand what the Germans at the next table are actually saying about you.

When the aliens arrive, one of the first things to send them reeling will be our inability to solve the language problem. We have atom bombs and flat-screen TVs, and we have even gone beyond the digital watch stage in our evolution. Yet still we struggle with silly books which tell you how to ask your Basque hotelier to run a bath for you at 70 degrees before polishing your batman's shoes.

Those of a romantic bent applaud the diversity of human babble. Different languages are more than different ways of encoding information, they say; they represent alternative modes of thinking, unique takes on reality. It must say something about the Irish character, for instance, that their old, odd language contains no distinct words for 'yes' and 'no'. Some of the tribal languages of New Guinea contain many words to describe different shades of green, but little concept of number. To lose a language – and only one in 10 is expected to survive the century – is a global tragedy.

Yes, yes, yes. But pretty and romantic and sociologically insightful as they are, foreign languages are an utter bore and get in the way. The answer is not to turn the world monoglot overnight (this will probably happen anyway in the next few hundred years), but to find some way of making language learning easier or translation quicker. As generations of taciturn teenagers and eager pensioners have found, language acquisition after the age of about seven takes an awful lot of hard work and time. Unless you are Dutch, of course, when apparently something like a Babel fish (perhaps dredged from the canals) is implanted into your ear as a child.

Fortunately, we have computers. They might not be able to teach us to speak, say, Spanish, but surely they can do the job for us? Yes, of course they can. Online, as I write, there is just the thing to perform this task, a natty little translation program which is enormously popular. Let's see if it works. Here we go:

Call up website.

Type *The practical upshot of all this is that if you stick a Babel fish in your ear you can instantly understand anything said to you in any form of language* into the website and tell it to turn this into Spanish.

Cut and paste the result into the translation box again and tell it to turn it back into English. Oh, for fun, make it take a detour through French and Greek on the way.

The end result is this: *The expert upshot all is that if you in stick a fish Babel in his hearing then you you include immediately all thing this you have you in all form of sole.*

It is a bit like taking a hundred-dollar bill into Mexico and working south through the *bureaux de change*. By the time you get to Patagonia it will be worth less than the price of a beer. An awful lot is, quite literally, lost in translation. And this can have terrible consequences. When a chance remark by Arthur Dent went winging its way across space and time via a wormhole, interpretation difficulties meant that his words sparked a terrible war that lasted centuries and killed millions. Grief on this scale is unlikely to ensue in, say, a French restaurant, but nonetheless, a machine that could take the pain away would be terribly useful. Sadly, however, building one is proving to be extremely difficult. The question is, why? Why is it proving so hard to build a machine that can make sense of what we say? After all, the average two-year-old can pick it up quite easily.

The technical term for what the Babel fish does, or would do, if it were a computer, is machine translation. MT has a history almost as long computers themselves. 'It started in the 1940s', says Professor Alon Lavie of the Language Technology Institute at Carnegie Mellon University in Pittsburgh. Machine translation, Lavie explains, 'was one of the first conceived applications for computers'. In the 1950s, 'toy' translators were built which looked promising. The first public demonstration was at Georgetown University, when 49 Russian sentences were translated into English using a 250-word vocabulary. Everyone got highly excited, until it was realized that beyond the 'My name is Sam' stage such translators failed miserably. In the 1960s, machine translation was recognized as a very hard problem. In 1964, a body called ALPAC – the Automatic Language Processing Advisory Committee – was set up by the US National Academy of Sciences to investigate the future of MT.

Two years later, ALPAC reported back with the conclusion that MT was slow and unreliable and that research to date had not yielded useful results. The flow of tax dollars into MT was promptly stopped and computerized language translation was put on something of a back burner.

Throughout the 1970s and 1980s academics continued to wrestle with the formidable challenges of interpreting our prattle. MT advances went hand-in-hand with voice recognition technology – an essential development if anything like the Babel fish is ever to become a reality. In the 1980s, DARPA – the inventors of the Internet – developed speech recognition programs and in the mid-1990s this technology became available over the counter, in products like IBM's ViaVoice.

Since then the market for a good MT system has mushroomed. The growth of the Internet and supra-national bodies like the EU and the UN, as well as the explosion in international trade and tourism, means that there is a huge demand for translation services. One response to this demand has been the creeping dominance of English as the global *lingua franca*. If current trends continue, then, in a thousand years' time English – or a bastardized version of it – may well have fully taken over. In fact, the inherent bastard nature of English has been its main strength. It is unusual in having so many distinct roots – including Greek, Latin, German and French. Also, while English is a hard language to learn well – having a very large vocabulary – its relatively simple grammar makes it fairly easy to achieve reasonable fluency compared to many tongues. Some other language may win the day – China's GDP is poised to overtake that of the USA in the early 2040s, so we may all be speaking Mandarin in a couple of centuries' time. But given the established dominance of English – in culture, commerce, tourism and academia – it is hard to see it losing its preeminence anytime soon.

Meanwhile machine translation researchers struggle on. According to Lavie, a key headache is ambiguity. 'Human languages are highly ambiguous, and differently in all languages. The ambiguity exists at all levels: lexical, syntactic, semantic, language-specific constructions and idioms.' In other words, the rules of grammar plus a dictionary are nowhere near enough because meaning is so heavily dependent on context. He calculates that a good MT system will require a lexicon of several hundred thousand words and about as many phrases. Plus translation rules. 'How do you acquire and construct a knowledge base that big that is (even mostly) correct and consistent?' he asks.

The answer seems to involve brute force, thousands of person-years of human expertise developing colossal word and phrase glossaries and regulations which then have to be digitized. Ideally, machine translation systems will take any language and generate a detailed, symbolic representation of meaning. This amounts to creating a universal language, an 'Interlingua', which could act as a mediator between any two real languages. 'Nice in theory', warns Lavie, 'extremely difficult in practice.' The go-between language has to represent unambiguously all the pieces of meaning that need to be translated and it has to be completely neutral, with no inherent bias to any real language. Too complex, and languages cannot be mapped accurately; too simple, and nuances fall through the gaps. Interlingua is just one of several angles computer scientists are trying. The newest tack for MT is the statistical approach. Here a program analyzes very large amounts of text – Canadian parliamentary proceedings logged in both French and English, say – to learn how key patterns and phrases interrelate.

Speech-to-speech MT is far more difficult than translating text. Spoken language is messy; it lacks helpful punctuation and it is full of 'erms' and 'ahs' and false starts as well as the horrors of dialect and differing pronunciation. Even building reliable speech recognition systems for one language has proved far more difficult than was once thought – try booking a cinema ticket or getting hold of an American telephone number using one of these infernal systems.

Lavie thinks that perfect MT is still a way off. 'My guess is that we are 5–10 years away from speech translation of really high quality', he says. But he is surprisingly optimistic that we will soon be able to buy a machine to provide rough-and-ready translation from the spoken word that we can plug into the ear – a true electronic Babel fish. 'It is mostly a hardware issue', he says. 'My guess is that by the time we get machine translation to perform well enough, getting it to run real time on a small device will be a non-issue.' In other words, the very globalization that is threatening to turn the world monoglot might just provide the economic and technological impetus for us to carry on speaking myriad tongues.

9

teleportation

I teleported home one night
With Ron and Sid and Meg.
Ron stole Meggie's heart away
And I got Sidney's leg.

*Song chanted by sceptical crowds outside the
Sirius Cybernetics Corporation Teleport Systems factory on
Happi-Werld III*

Getting from A to B has, in theory, never been simpler. In the United Kingdom alone, you can choose from more than 200 models of automobile with a top speed of above 225 kph – over a fifth of the speed of sound. Airliners can whisk us half-way round the world in a day, a journey that took weeks by steamship. The internal combustion engine has given human-kind the gift of speed, arguably the first truly new experience of the 20th century.

Yet, despite our fast cars and faster planes, travel is, as it always has been, a pain. It takes hours, whether you are going to Singapore or Southend. Buses occupy a parallel universe that almost but never quite coincides with your own. British train timetables are works of baroque fiction that any sensible publisher would reject. Every nation, from richest to poorest, has its own transport grief. Americans live in a vast country and seem to need to get about a lot, but in a cruel irony they are punished with the planet's most terrible airports and dreadful cars. In Los Angeles they tore down the tramways and built a freeway system that was the envy of the world, yet which at peak times slows the city to the speed of a trotting horse. In developing countries, getting around usually means walking, relying on some sort of animal or, if you are lucky, crowding onto a deathtrap bus and venturing onto roads that lead straight to shrieking madness. And as for Europe – if you have ever driven across Naples at six o'clock in the evening you will have had a glimpse of eternal damnation itself.

What we need is a way to get us where we are going, instantly, without pollution or fuss. This is where the teleport machine comes in. The transporter of *Star Trek* fame is still, sadly, a fiction, but unlike time travel, it looks likely that some sort of teleportation is possible even with today's limited technology. There is no need to make tame black holes, or to squash Jupiter. All that's called for are some clever tricks with lasers and mirrors. We are a long way from knowing how to teleport Arthur and Ford out of a Disaster Area concert, but we might be close to teleporting a virus or even a bacterium, which would be astounding enough. This would require a truly extraordinary exploitation of the strangeness of quantum physics. And like time travel, teleportation opens up a philosophical can of worms.

The teleports invented by science fiction writers – one of the many devices first proposed in fiction, including the space rocket and the atomic bomb – usually work by disintegrating an object or person in one place and making a perfect replica somewhere else. It is not clear how this happens. Sometimes the idea is that some sort of scanning occurs, and the scanned information, rather than the atoms and molecules themselves, is sent from A to B. At B, this stream of data is used to recreate the original object. Other times, as with the *Enterprise*'s transporter device, it appears that something of the physical essence – the actual atoms – of the teleportee is moved from A to B, as well as the information needed to remake him or her. When Captain Kirk is beamed down to a planet's surface (or on rare occasions into the vacuum of space), there appears to be no bucket of atoms at the other end from which his body can be recomposed.

Either way, the classical teleport is analogous to a fax machine. When you send a facsimile, you are not sending the letter itself. You are sending the *idea* of the letter, which the recipient's machine then recreates. The *Star Trek* transporter also sends the pulverized paper on which the original was typed, but the principle is the same – although, unlike a fax, the original letter ceases to exist.

Is a faxed letter the same as the original? If a letter is just a way of conveying information, then yes. As long as words, font and even the colour of the paper have been copied accurately, there should be no difference. But most people would be unhappy with this idea. For a start, there are now two letters, and only one can be original. A faxed document does not always have the same legal status as the original; when signatures are concerned, there seems to be more to it than the simple information conveyed in a handwritten squiggle. This difference reflects our philosophical intuition; there was an outcry in late 2003 when it transpired that condolence letters sent to bereaved American families who had lost relatives in the Iraq war had been signed, not by Defense Secretary Donald Rumsfeld, but by a machine which produced a replica of his signature. A signature is given authenticity by its history. It must have been formed by the fingers of its originator; his or her skin must have brushed across the paper upon which it is written.

The problems with teleportation, technical and philosophical, stem from the copying process. Take the simplest scenario. You step into the teleport machine – perhaps a booth or chamber much like the one in the archetypal teleport movie, *The Fly*, released in 1958 and remade in 1986, starring Jeff Goldblum.

Inside the booth some sort of beam, be it a laser, X-rays or whatever, analyzes your body and records the position of every atom in you. This is a big job, as the average human body contains some 7,000,000,000,000,000,000,000,000,000 atoms, but it is possible. Counting, identifying and plotting the position of each atom is way beyond the most powerful computers we possess, but that may not always be so. Let's assume that it can be done. In being analyzed, your body is destroyed. Hopefully this is painless and more or less instantaneous – it always is in the movies – but again, no matter. All the information on the nature and whereabouts of those atoms is then transmitted, maybe down an electric cable, maybe by radio waves, maybe by some sort of quantum process, to another chamber: the receiving teleport station. There, your body is reassembled atom by atom – again this is usually assumed to be more or less instantaneous as the alternative would be gruesome – much like a three-dimensional facsimile. Then what?

This is where the problems, and the fun, start. In science fiction, the problems tend to hinge on two possible horrific scenarios. There is the one alluded to in the *Hitchhiker's* ditty at the start of this chapter – the *Fly* scenario. Here, some foreign object is unwittingly allowed to enter the teleport booth along with the intended traveller. During the scanning process the properties of the foreign body are somehow scrambled and merged with those of the unfortunate in the machine, and when the legitimate passenger is reassembled, an essence of the stowaway is reassembled with them. Thus, in the movie, when a fly enters the chamber with Seth Brundle, and he is zapped across the laboratory, his DNA and the fly's are merged, with distressing results for all.

The second horror scenario concerns a malfunction in the teleport machine, which means that it cannot make a perfect copy of the original – much like a glitch in a fax machine meaning the letters come out wrong, or that the lines are mangled. In the first *Star Trek* movie, the grim consequences of such a malfunction are made clear, when two crew members are teleported inside-out onto the transporter pad of the

Enterprise (the same happens to a monkey in *The Fly*). And if not fatal, even a well-oiled teleport machine is no picnic to use. In the first instalment of the *Hitchhiker's* series, Ford Prefect and Arthur Dent use a teleportation device to escape from the Earth just as it is about to be demolished by the Vogons, ironically, in the name of faster transport. The shock to the system, Ford explains to Arthur, can only be mitigated by copious quantities of salt, protein and muscle relaxant. So if you are going to be teleported, have a pint of beer and a packet of peanuts.

The deepest problem with teleportation does not depend on a malfunction. Remember how the machine works. You walk in; you are copied, and remade somewhere else. In the process the original 'you' is destroyed. Does this matter? On one level, no. Imagine that the object to be teleported is not a human being but, say, an old car. You put the car into the machine and *shazam!* It appears a kilometre away. Every single atom of the 'new' car is identical to that in the old one – an iron atom is an iron atom, after all. So are the positions of every single atom and the quantum state of every sub-atomic particle. You have a car that looks, feels, smells and drives just like the original. Every scuff on the paintwork is there, every dog hair on the back seat, every chip in the windscreen and rust pit on the bumper. You walk up to the teleported machine and examine it, recognizing every dent, every oil stain. It is, in all important ways, identical to the original car. In fact, it is more than just a very good replica – it is the actual car. According to the physicist Brian Greene, 'if two particles of the same species are in the same quantum state ... the laws of quantum mechanics ensure that they are indistinguishable, not just in practice but in principle'. In other words, if your old car were teleported, there would be no way, even in theory, for you to ascertain that this had been done. You would be looking at, in every sense, your old car. Except that something niggling away at the back of your mind tells you that it is not.

Take a real-world example that deals with the same philosophical issue without recourse to teleport machines. A few years ago, I read a story in a classic car magazine about how an enthusiast rescued a very rare old Rover saloon he discovered rotting in a field. The car was in a terrible state, but as there were only about half a dozen like it in the world he considered it worth saving. He went through the car deciding what needed to be done. The chassis was too rusty to be salvaged, so he replaced it with one from a similar model. Most of the body panels were shot too, so he swapped them for carefully crafted sheets of aluminium, cut out by hand. The engine? No hope there, but he sourced a replacement from a similar car. It was the same story for the transmission, the axles, the wheels and the interior. No matter, the enthusiast rebuilt the car. In fact, just about the only original part on the new car was the badge – and this was so valuable that a copy was made and the original kept safely inside. The magazine presented this as a heartwarming story of an old jalopy made new, but this is not what had happened. A *replica* had been made, not a *restoration*. When some readers complained of this on the letters page, however, they were shot down by their fellow enthusiasts. To give an even more prosaic example: I have a wonderful old axe in the shed. It's had three new handles and two new heads.

Most of us believe there is more to an object than the atoms of which it is made. Even if these atoms are identical in every respect, there is something about a teleported object that is just not the same. The problem gets a whole lot worse if you are talking about a human instead of a car. If teleportation involves your body being destroyed in one place and recreated somewhere else, what is it that is teleported? You, or a copy? Because surely there is no difference between teleportation and stepping inside that booth, putting a gun to your head, and pulling the trigger. Teleportation kills its subject. Just because a person looking like you, claiming to be you and indeed believing that he or she *is* you, with all your memories in his or her head, including the recent one of entering the teleport booth, walks out the other end does not change this. As far as the original *you* is concerned, life ended when the button was pressed. The new person is an impostor.

To see the logic in this argument, imagine that instead of re-creating the teleported person instantly, the information needed to make a copy was stored for a while, say 100 years. Say the machine storing the data was itself destroyed at some point, but that a new machine was built and the data recovered from the old, broken, teleporter. Say this takes place centuries hence. The original you has been dead for years. When the delayed teleportation process is finally completed, most people would accept that this is merely the creation of a very good replica. Or consider another variation. Something goes wrong with the teleport machine. It fails to kill the original you in the scanning procedure (in fact this is probably not allowed, for various complicated reasons that will be explained later). You survive the scanning, but the copy is still made. There are now two people claiming to be you – two who fervently and genuinely believe they *are* you. Remember, everything about you, including all your memories, is recreated in the copy. Say the machine malfunctions more seriously. You pay Teleport Inc. £1000 to be zapped from London to New York. One copy is made in New York, but the original you survives and the data to make the copy is mistakenly also sent to booths in Munich, Sydney and Montreal. There are now five of you. No problem, says Teleport Inc. We will, er, dispose of the four unwanted copies, including the original you, and let the one in New York walk away. No extra charge, step this way sir, it will be extremely quick and painless ....

Teleporting someone in this way is clearly murder. Or is it? Because now we get to the strange and bothersome business of trying to pin down what we mean by identity and authenticity. As many people have pointed out, something like teleportation – in the sense of you being destroyed and replaced with a replica – occurs, quite naturally, to each of us all through our lives. The body 'I' now inhabit is not the same one that I inhabited 20 years ago. For a start, it is older and a good bit fatter and wrinklier. But the change is more profound than that. My cells are dying and being replaced all the time. The chances are that a large proportion of my eighty or so octillion atoms are not the same as those that comprised me 10, 20, 30 years ago. Forty years ago I was about 18 inches long and weighed eight pounds or so. I am the same person I was then, but I now weigh 25 times more – physically, I am not even a bad copy. I am new atoms. It is true that most of the cells in my brain – the bit of me where 'me' can be said to live – are original, but many are not. We now know that adult brains regenerate cells throughout their lives. In effect, the young me has been teleported into the future as a flabbier copy, with new atoms, and not even an approximation of the old shape. But no one would dispute that it was still me, least of all ... me. If this is the case, then surely teleportation is *not* murder?

The issue is destruction and recreation – does the essence of the individual survive? No one has yet performed a brain transplant (which should really be called a body transplant), but people have transplanted bits of brains into people. Injections of foetal brain cells have been used to treat Huntington's Disease and other degenerative conditions. It is all highly experimental and early days, but this sort of medicine raises interesting questions. Say your brain is damaged by accident or illness. Soon neurosurgeons might be able to repair that damage, growing genetically compatible replacement neurons in the lab, or even, as has been suggested, artificial neurons on integrated circuits, and stitching them into your broken grey matter. At what point can we say that the original brain was so damaged that we are talking about, or to, a new person? When has the doctor made a new classic car, rather than repaired the old one? As long as the surgery is successful and, to an outside observer at least, the patient seems to have retained his or her memories, personality and so on, the answer is 'never'.

You don't even need to have brain surgery to enter this dilemma. Every time you fall asleep your brain switches off – not totally, but enough to interrupt the sense of continuous conscious existence that you have throughout the day. This does not matter, because eight hours later you wake up and the brain starts again. In fact, every second – every tenth of a second – awake or not, the electrochemical state of your brain changes. The brain you have now, this instant, is not the same brain you had five minutes ago, certainly not on the quantum level. What it retains is the *feeling* that it is the same brain with all the same memories. You can be said to be dying and coming back to life every second, without a moment's thought.

So what is the essence of existence that is so challenged by teleportation? Memories? Nope, because these can be preserved. The atoms you are made of? This cannot be so. Every atom of hydrogen in the Universe is profoundly and utterly the same as all the others. Is it continuity of existence of these atoms? No again, as this does not happen in the growing body. Is it the fact that the atoms occupy the same space? No again, because if that were the case you would be a new person every time you moved. So why should we be unhappy about teleportation? I for one would never get into a teleport machine, but Brian Greene would. In his book *The Fabric of the Cosmos* he states:

> Would the person who steps out of the receiving chamber be the same as the one who stepped into the teleporter? Personally, I think so ... to my way of thinking, a living being whose constituent atoms and molecules are in exactly the same quantum state as mine is me. Even if the 'original' me still existed after the 'copy' had been made, I (we) would say without hesitation that each was me. We'd

be of the same mind – literally – in asserting that neither would have priority over the other'.

Some would disagree. Those who believe that there is more to human identity than their physical state might say that this essence – call it a soul – is destroyed by the teleportation process. In the same way, classic car enthusiasts and dealers debate the nature of the 'soul' of an old automobile. Interestingly, there is a fairly strict definition, involving original bits of chassis, engine blocks and so on, which have to be original for the car to be the 'same'. These rules are in place to prevent unscrupulous mechanics passing off copies of valuable old cars as originals; the restored car I described earlier would fail the test. If only it were so simple with living beings.

Teleportation then, like time travel, has its philosophical quandaries. For what it is worth, I am not a dualist and I do not believe in souls and other spookery. But even so I am not sure I buy Greene's argument. As another Happi-Werld song goes, 'if you have to take me apart to get there ...'

There is one big difference between teleportation and time travel. And that is that teleportation is actually fairly easy. Some of the weirder aspects of quantum physics allow us to send an object from one side of the room to the other in an instant – or at least massively faster than the speed of light. This seems odd, because at first reading quantum physics specifically precludes teleportation. Heisenberg's Uncertainty Principle states that it is impossible to know both the position of an object and its momentum at the same time. If you scan a microbe, a person or a car you could get each atom's position right, but you could not tell its velocity as well. In fact, says the principle, this goes for any pair of attributes. You cannot measure the precise quantum state of something even as simple as an electron. (On the *Enterprise*, a 'Heisenberg Compensator' plumbed into the transporter machinery takes care of this niggling inconvenience.)

Nonetheless in 1993, a team of six scientists from the USA, Canada and Israel, led by Charles Bennett of IBM, came up with a theoretical teleport machine that, far from being scotched by the quantum world's oddities, actually relied on them. Their thought experiment involved a quantum property called entanglement. This is the name of the communication that seems to occur between pairs of quantum objects such as photons or electrons. Being entangled means that although any measured property of an object, such as the polarization of a particular photon, is random and seemingly unrelated to anything else in the Universe, its entangled partner will have the same property – even if it is a million light years away. Entanglement is often called the EPR effect, after Einstein and Rosen and their colleague Boris Podolsky. In 1935, these three deduced how entanglement works over large distances. It was Einstein, never entirely comfortable with the quantum world, who coined the phrase 'spooky action at a distance' to describe the effect.

It looks like entanglement would be the perfect way to transmit information over large distances faster than light; however, at first glance this appears not to be so. We can find out that distant photon B is polarized vertically by measuring the polarization of local photon A. But measuring photon A changes its state, and we have no way of knowing what its polarization was before. We know only that once the measurement has been made, both photons are in the same state.

There is a way round this, which allows us to transmit an object's actual quantum state, as well as information, from one side of the room to another in an instant. The 1993 theoretical experiment by the Bennett team was carried out in 1997, when a group of physicists led by Anton Zeilinger at the University of Innsbruck used quantum entanglement to teleport a single photon at sub-light speed.

Their solution is to use another pair of entangled photons as messengers to carry the message from photon A – the one you want to teleport – to the photon at the other end of the line. Touching the photon – measuring it – causes the whole thing to collapse, but if you use entanglement to do the measuring it is as if you are handling the photons with disinfected gloves, letting them remain in their pristine quantum state. Here is how it works. Take three photons: A B, and C. Entangle B and C. Put B next to the outward teleport booth, and carry C to the receiving end. Now – and this is what Bennett and his colleagues showed was possible back in 1993 – you can measure certain things about photons A and B without directly affecting photon A. For instance, you can measure whether A and B have the same spin without actually needing to find out what that spin is.

Now you know something about A in relation to B, and you know that B is entangled with C, so you can transmit the precise quantum state of photon A to the receiving end of the teleporter. The teleport machine can now use the information at its disposal to calculate the original state of photon A and replicate it in photon C – which is now effectively photon A post-teleportation.

In the process, the original photon A – its quantum state before all this malarkey – is destroyed. When teleporting your photon, you destroy the original. Zeilinger and his team did this in 1997, using a laser, beam-splitters, a series of polarizers and mirrors, and a kind of crystal that creates entangled pairs of photons when a laser passes through it. A year later, scientists in Pasadena and Denmark used the same technique to show it was possible to teleport a whole beam of light across the lab – a feat achieved by Australian scientist Ping Koy Lam in 2002. 'What we have demonstrated here is that we can take billions of photons, destroy them simultaneously, and then recreate them in another place', Dr Lam said at the time.

Is this actually teleporting? After all, it is not the photon itself you have zapped across the room, just its quantum state, which you have then imparted to another photon. According to physicists, the quantum state of an object *is* the object. As Zeilinger wrote in *Scientific American*, 'Particles of the same type in the same quantum state are indistinguishable, even in principle ... identity cannot mean more than this: being the same in all properties'. And, he points out, this is a far more profound way of getting stuff around than faxing. In a fax, you end up with two letters. With quantum teleportation the original is always destroyed. And with faxing it is fairly obvious which is the original and which is the copy. With quantum teleportation the 'copy' is identical to the original in the deepest way. Physicists H. Jeff Kimble and Steven Van Enk wrote in a commentary piece in *Nature* on the subject that teleportation can be characterized as the 'disembodied transport of quantum states'.

Of course, photons are not cars, cats, flies or humans. Light beams – ghostly objects without mass – are one thing. The world of real hard stuff is quite another. Could this technique extend to include larger objects?

The answer appears to be 'yes', at least in theory, and with the caveat that colossal amounts of computation will be required to teleport any object that we could see. In 2003, scientists in France demonstrated entanglement in pairs of atoms – hulking great behemoths in quantum terms. And if you can entangle atoms, you can teleport them, something duly done in 2004 by Rainer Blatt and David Wineland. Blatt, of the University of Innsbruck, and Wineland, of the National Institute of Standards and Technology in Colorado, working independently, used the same technique applied to photons. First, they entangled two calcium ions: B and C. Next, they made a third atom, A, bearing the state to be teleported. Then they entangled A and B. They measured the state of both and sent the result to C. This transformed the quantum state of atom C into that of A, destroying A's original state. Blatt and Wineland had teleported A to C's location.

If you can entangle and teleport an atom, you can entangle and teleport a molecule. And if you can do that, why not a virus? No reason, it seems, except that to teleport an object so large would involve transmitting and decoding quadrillions of bits of information from one teleport booth to the other, and then using this information to impart quantum states to the same number of quantum objects in the receiver booth – the protons, electrons, neutrons and so on, from which you are going to build your replica.

It would seem that the *Star Trek* transporter device – the most famous teleport in fictional history – cannot use this method to get crew from ship to shore, as it is suggested that the actual material, as opposed to just the quantum state of the material, which makes up the crew members is teleported. Somehow, Captain Kirk's actual atoms are taken and zapped where they are wanted. How does this take place? It is not explained, except that the atoms (and the information needed to recon-stitute the crewman's body) are held in some sort of 'buffer' device. The suggestion is that somehow the object to be teleported is quite fundamentally 'dematerialized' – its atoms smashed into free quarks and electrons – before being zapped through space (and solid and fluid objects such as the hull of the *Enterprise* and the planet's atmosphere). To do this – just the dematerializing bit – would require quite a lot of energy, several million times that released by a large thermonuclear explosion.

We may never see a human teleported. If we could build the machine, who in their right mind would get in it? But it is con-ceivable that we will, before the century is out, see objects as large as microbes moved by that strange process we call en-tanglement. In the more immediate future, quantum tele-portation has been mooted as a way of carrying information from one part of a computer processor to another – in quan-tum wiring. Next to all this, Einstein's tardy clocks and warped space–time look reassuringly old-fashioned.

10

meat with a clean conscience

May I urge you to consider my liver? It must be very rich and tender by now, I've been force-feeding myself for months.

Milliways' Dish of the Day

That's cool. We'll meet the meat.

Zaphod Beeblebrox

We are a long way off breeding an animal that both wishes to be eaten and which is capable of saying so, clearly and distinctly. Yet there is growing unease about the way we seem to be tinkering with our food, as the lurid GM scare stories emanating from the UK's tabloid press illustrate. The increasing popularity of organic and free-range meats, in Europe at least, is a sign that we are becoming disaffected with the monstrosities that are being visited upon farm animals in the name of increasing yields and cutting costs. Even without genetic modification, there are double-muscled cows and supertender sheep. We have bred turkeys too grotesquely big to mate (demanding the creation of a whole new career – professional turkey masturbator) and chickens that reach adult weight in weeks.

In America, home of perhaps the most degraded palate on the planet, there has been little controversy to date on genetically modified food. Americans, reading the headlines in the

European press, are apt to ask, 'Hey, what is this problem with General Motors?' – perhaps not surprising when even normal, DNA-intact, American fare is truly horrible. It is the land of enormous shiny red apples that taste of cotton wool, lurid pink steaks given not a moment's pause for reflection and flavour development, and, of course, the horror that is 'lite'. Artificial spreads and oils, cheeses and candies pack every supermarket shelf from sea to shining sea. These substances are all lower in fat than their natural equivalents, but they tend to be much higher in sugar and salt to compensate, which may have some bearing on the current American waistline.

In Europe, things are different. Well, most of Europe. British food can still be almost as horrible as American, yet even here there has been a widespread rejection of transgenic produce across all sectors of society. The situation in continental Europe is far more passionate. In France there have been physical attacks on the manifestations of US food culture, including the bombing of McDonald's restaurants. In Italy the thriving slow-food movement is a rejection both of vile transatlantic cuisine and of something more fundamental – the idea that food is essentially fuel, fodder that is fine to consume on the hoof instead of round a dinner table. The European reaction is as much against the whole American ethos of hard work, long hours and conspicuous consumption, as it is generated by fears that vat-synthesized victuals are going to kill us.

With genetic modification (to date mainly of fruit and vege-tables), the two food cultures have met like colliding waves – causing a foamy crash of controversy and a sea of confusion. Whatever the antis maintain, there is no evidence that genetic modification is harmful to health. Indeed, high-vitamin or vac-cine-containing varieties might be beneficial and drought- or salt-resistant breeds could become essential. On its environ-mental impact, the conclusions are more mixed. A recent UK government-funded study of GM crops, for example, came up with some confusing results. As far as wildflowers and fauna were concerned, some crops appeared to encourage bio-diversity, while others reduced it.

The biggest anti-GM story to date occurred in 1998 when Arpad Pusztai, a Hungarian émigré plant scientist working at the Rowett Institute outside Aberdeen, conducted a series of experiments in which he fed potatoes containing a snowdrop gene to a group of laboratory rats. When he found that some of the rats became ill, there was much brouhaha, and eventually in October 1999, his results were published in *The Lancet*, one of the world's most respected medical journals. Pusztai's critics – supporters of GM food – rounded on him, claiming that his paper was riddled with serious statistical errors and that he had shown absolutely nothing of interest regarding the relationship between transgenic vegetables and health.

Since then debate over GM food has become somewhat sterile. We are, after all, talking about fields of wheat and maize here. No one has died, no one has been injured. You can call them Frankenfoods all you like, but even a strawberry with fish genes does not walk around with a bolt through it neck scaring the villagers. For true Frankenfoods we must look to the (near) future.

It will probably be impossible, not to say undesirable, to create an animal that articulates its desire to be consumed. But what about one which doesn't mind being eaten? Even the most dedicated carnivores must admit that meat-eating cannot be carried on without a degree of suffering. From battery hens to *foie gras*, the end result may be tasty but somewhere along the line some poor critter will probably have had a distinctly miserable time of it. Can this be avoided? Would it be possible to eat meat with a totally clear conscience?

One man seems to think so. Morris Benjaminson has been working on a NASA-funded project to grow meat in a lab dish. The idea is to provide tasty and nutritious food for astronauts on long-distance missions, without the time, space and hassle of, say, keeping chickens or fish on board. Writes Benjaminson: 'The havoc and discord typical of early pioneering sea voyages punctuated by bad water, weevily hard tack, rancid salt pork and rampant scurvy, is not desirable and not necessary in the Space age'.

Today's astronauts do not have to put up with weevily bis-cuits, let alone rancid salt pork, but their diet can be pretty monotonous. Along with all the other privations, being in space for a long time apparently has an incredibly deleterious effect on the taste buds, to the extent that already bland space food ends up tasting like semolina after a few weeks. To combat this the Russians soon learned to use copious amounts of chilli and garlic to spice up their food, so much so that new cosmonauts visiting old hands aboard the Mir space station often used to complain the diet was almost inedible – and that the first sniff of Mir through the airlock left them reeling. Going to Mars or beyond will be much worse. There will be no supply flights, so everything will either have to be carried from Earth or grown on board. Astronauts could, of course, survive per-fectly well on a strictly vegetarian diet, and plenty of people do here on Earth. But sitting in a tin can for nine months will surely be bad enough without having to subsist on alfalfa and lentils.

So far, success has been limited in the artificial meat project. Benjaminson's New York-based team have taken gobbets of muscle from goldfish (freshly killed) and persuaded these to grow into mini-fillets an inch or so across in a saucer of nourish-ing serum. When they tried the same trick with chicken the 'meat' grew by about 14%. It only stopped growing (and started rotting) when the area of dividing cells was too far from the supply of nutrient carried through the flesh by blood ves-sels. In the goldfish experiment, the team washed the result, and gave it a quick dip in olive oil, garlic lemon and butter. Then they fried it and showed it to colleagues from other departments. It looked and smelt like fish, but unfortunately health and safety rules (so they claimed) meant that no one was allowed to find out whether it actually tasted like fish. Still, quite a promising start.

To grow a slab of meat big enough to feed a hungry astronaut, blood-vessel growth as well as muscle bulk will be necessary. Efforts are now under way to stimulate capillaries using electricity.

Tissue engineer Vladimir Mironov at the University of South Carolina suggests an alternative to raising fibrous muscle tissue in the lab, with all the irksome plumbing issues this entails. He thinks it may be simpler to culture animal protein in amorphous blobs and then process this proteinaceous mulch into artificial meat – rather like chicken nuggets. The simplest food to grow is fish, he says, but 'chicken is nice'. His dream is that we will one day be able to grow and cook a fresh sausage overnight in a machine much like a breadmaker.

Growing steaks in a test tube is always going to be hard. Getting protein to form is one thing; replicating the texture and taste of a real steak is quite another. When you chew a chicken drumstick, for example, you are chomping into thousands of elastic muscle fibres whose previous job was to make the chicken's legs work and move it around. And the freer and wilder the animal, the more it moved when it was alive, the better-tasting its meat. That is why free-range chicken tastes of chicken and battery chicken tastes of plastic, and nuggets of reconstituted mechanically recovered chicken taste of sewage. Could we replicate the real-meat taste and still avoid cruelty?

Now we are nudging the realms of science fiction, but it may be possible, one day, to genetically engineer certain breeds of livestock in such a way that they still produce meat in a normal way yet suffer absolutely no pain, stress or anxiety while being reared or killed. To do this, we would need to create an animal effectively without a brain – or at least without the parts of its brain that control awareness, pain and so on. As we understand the genetics behind embryo development, pain and memory we might create new types of animal which move, feed and excrete just like a normal cow, say, yet which have no conscious awareness either of the world of their own existence. Some have suggested breeding livestock with no sentience whatsoever, with just the 'primitive' brain stem intact to control basic motor functions, hunger and so on. This raises interesting ethical questions. Would it be immoral to physically ill-treat a creature that could feel no pain? How far would we go? There seems to be no reason why biotechnology could not come up with living versions of Benjaminson's goldfish – breathing, moving, growing insensate modified mammals or birds from which we could harvest an endless supply of protein. This sounds disgusting, but then that is the only word to describe what goes on in a chicken or turkey farm today.

There are currently nearly six and a half billion people on the planet, and that number will grow, probably to eight billion or more, before the population stabilizes. That is a lot of mouths to feed. The raising of livestock is energy- and land-intensive, one of the main arguments put forward by vegetarians for the wholesale abandonment of meat-eating by humankind. But it is highly probable that many people – billions of people and growing – will want to carry on eating meat. Thirty years ago we had the 'green revolution', when new varieties of staple crops like rice enabled farmers in much of the Third World to increase yields massively.

The coming years could see a 'red revolution', if the yuk factor can be kept at bay, in which animals genetically modified for efficiency replace traditional breeds (most of which are products of centuries of selective breeding anyway). We have already seen the 'Frankenfish', a gene-enhanced salmon which can grow to ten times the size of its natural cousins. By twiddling around with animal DNA, it has proved to be possible to do more than to enhance yields. Cows, for instance, can be programmed to produce useful pharmaceutical chemicals in their milk. Most bizarrely (to my mind), a firm in Canada called Nexia Biotechnologies Inc. came up with the bright idea of splicing spider genes into goat DNA (of all things) to produce animals which make strong silk threads – BioSteel® – in their udders. This has led to what must be one of the strangest new buzz terms in biological history, 'transgenic goat technology'. All this before we start worrying about the relationship between cruelty and our dinner.

There is something, even the most dedicated meat-eater will admit, that is inherently unpleasant about shooting a sentient animal, then carving it up and eating its buttocks. Most of us just shrug our shoulders and live with it. Creatures have been killing and eating each other, after all, ever since there were creatures. But a steak grown in the lab? Even one that *can't* talk to you? A godforsaken salad may be preferable.

11

the total perspective vortex

The Universe, the whole, infinite Universe. The infinite suns, the infinite distances between them, and yourself an invisible dot on an invisible dot, infinitely small.

Pizpot Gargravarr

Perhaps the cruellest device ever imagined, the Total Perspective Vortex at first sounds not too bad. Rather like water torture, it seems far preferable to the stuff involving pliers, electricity, pokers and racks.

But to any aficionado of these more medieval techniques, the Vortex would appear like the inner circle of hell as compared with the North Circular on a Friday afternoon. The latter can make you lose the will to live, but only the former can tear your soul apart and make you wish that you had never been born and that the Universe had never been created to contain you.

The idea is simple. Every object in the cosmos is influenced, even to the smallest degree, by every other object, force and event; so in principle it should be possible to extrapolate the whole Universe from, say, a small piece of fairy cake. The person, the *Guide* tells us, who first hit upon this surreal and, as it turned out, infinitely psychotic idea, was a man, or thing (his species is not given) called Trin Tragula, who was fed up with his wife.

Tragula was a dreamer, a thinker, a speculative philosopher or, as his wife would have it, an idiot. 'Have a sense of proportion', she would tell him, perhaps 38 times a day. So he built the Total Perspective Vortex, in which the entire Universe and all its works were extracted and extrapolated from the summation of movements and machinations of every force and particle in existence – onto a small piece of matter, in this case, a chunk of cake. At the other end of the machine was his wife. When Tragula turned the Vortex on, she was confronted with the entire enormity of creation and – and this is the important part – herself in relation to it. This is clearly more than any sentient being can cope with. After a scream of horror her soul was fried. It was the ultimate punishment.

This is what was in store for Zaphod Beeblebrox, but fortunately for him he has a man on his side who had been good enough to create an entire universe for him. Thus, when confronted with the enormity of creation (and himself in relation to it), Beeblebrox can only conclude that his previously high estimation of himself was more than justified.

We now know that every atom in our bodies is or can be entangled with objects that might be billions of light years away. The Universe was, after all, once very small – and everything within it was once crammed together like peas in a pod. Everything might be linked in what Michio 'multiverse' Kaku calls a 'cosmic quantum web'.

Nonetheless, creating a Vortex might be harder than it first appears. For although every single object, force and event in the Universe undoubtedly impacts and impinges on every other object, force and event, extrapolating the whole from one of its parts may turn out to be impossible.

There is the tricksy nature of quantum mechanics, for example. Heisenberg's Uncertainty Principle states that it is impossible to know, at the same time, and with total precision, everything about a quantum system, even if we measure it with the finest instruments imaginable. Werner Heisenberg pointed out that with an electron, say, there is an infinite amount of uncertainty about its position. But this is only the case if the said electron is free to move around in space. If the electron is confined to a box, we have a much more precise idea about its position. So as long as we know exactly where the electron is, it is possible to say everything useful we can about it, right?

Er, no. Heisenberg showed that if we have a highly localized position wave function to describe the electron (or, in plain English, if we know exactly where it is), then there is a concomitant spread in its momentum wave function. So there will always be a large degree of uncertainty about the momentum and velocity of a localized electron. Even if you nail the blighter down, you will never know what it is up to. It is impossible to know the precise location and velocity of an electron, or any other sub-atomic particle. This would appear to preclude the construction of a machine that can infer everything from a small piece of fairy cake. Of course, there might be a way round this. The same quantum fuzziness that appears to rule out a Vortex also appears to banjax a teleport machine (see Chapter 9) but scientists are giving it a jolly good go.

Building a Total Perspective Vortex might be tricky, but the loophole that lets Zaphod escape with his soul intact may be easier to pull off.

A few years ago an intriguing idea appeared in *New Scientist* magazine. A variation on the planetarium hypothesis that we met in Chapter 2, it stated that we are statistically overwhelmingly likely to be living in a computer-generated universe, just like the one into which Zaphod crawls when he leaves Zarniwoop's office. It could be dubbed the Matrix theory, after the science fiction film of that name. The premise of *The Matrix* is this: at some point in the future, human computer scientists manage to build a machine, the eponymous Matrix, so powerful that it can model an entire world, complete with electronic human avatars, in its processors. The world of the Matrix and the bodies of its human inhabitants are no more than figments of a machine's imagination.

According to Nick Bostrom of Yale University, it is inevitable that one day our machines will be capable of doing this. And it is equally inevitable that we will try to build a virtual reality universe – and succeed. And – this is the really important bit – our successors will do this not just once, but countless times. Now, he argues, look around you. You see a universe. Knowing that if everything is as it seems we could one day build countless simulacra universes, what are the chances that everything really *is* as it seems? Very low. Because the probability is that, rather than living in the unique, genuine progenitor Universe – what Bostrom calls the 'original history' – we are living in one of the countless simulations. We are no more than electronic avatars, just like those in *The Matrix*. It sounds bonkers; it probably is bonkers; but it's intriguing.

The Matrix hypothesis depends upon it being possible to build computers that are conscious – or that at least mimic what we think of as consciousness. Many computer experts have their doubts about this, pointing out that our most powerful machines to date may be calculators *par excellence*, but are no more sentient than a lump of granite. Bostrom counters that true machine intelligence is no big deal. Hans Moravec, a visionary roboticist at Carnegie Mellon University in Pittsburgh, has done a back-of-the-envelope calculation and worked out how fast a computer would need to be to emulate a mind. This turns out to be about 10^{14} operations a second. At the moment, IBM's finest can manage about one-tenth of that speed, but remember Moore's Law: we only need another four doublings, six years or so, before Moravec's boundary is crossed. Assuming we can solve the riddle of what consciousness actually is, and how to simulate it in a machine, we should have computers powerful enough to do so by the mid-2010s.

Then, according to Bostrom, technologically it is downhill all the way. Simulating an entire universe would not be strictly necessary – it would just need to be good enough to convince its 'inhabitants'. There would be no point filling in each and every detail – only when somebody decided to look at a distant astronomical 'object', for instance, would the machine colour in its physical properties.

This is a disturbing idea, for it means that our simulated consciousnesses are at the mercy of beings whose intents and morality we can only guess at. Maybe they enjoy making us suffer. Maybe the best way of surviving such a Matrix-world is to stand out in some way. If we are all in a great big version of a computer game like *The Sims*, then our purpose in life could be to entertain our masters. Be glamorous, do something dramatic, like become a dictator or a pop star or a princess.

It is the paranoiac's perfect fantasy. Things are not only not as they seem, but far, far worse. Everything is an illusion, nothing is real. It is all, in short, a set-up. Of course, no hypothesis is worth it's salt unless it can be tested. And somewhat surprisingly this may not be such a long shot.

According to Bostrom, if the Universe is a simulation, it may be possible to spot the joins where the programmers did not do a perfect job – real-life 'disturbances in the Matrix'. In the movie *The Truman Show*, which features essentially a low-tech version of a simulated universe – a film set in a city-sized studio – our hero suspects something is up and decides to test his hypothesis. He travels furiously round the block, not following his normal route to work, and notices the startled reaction of his fellow townsfolk, who scurry away in confusion. If our Universe is not real there might be similar discrepancies. What would we look for? Well, we might look for some bizarre irregularities at the sub-microscopic level, which the computer would probably find hardest to model with complete consistency. And irregularities, somewhat worryingly, are exactly what we see. At the most extreme scales, physics does not add up. It has so far proved impossible to reconcile quantum laws with relativity, for example. Physics is trying hard to find a Grand Unified Theory of everything that will mop up the mess, but it hasn't got there yet. Trying to spot these sorts of inconsistencies might also be a bad move. If we *are* living in an illusion, then the moment we collectively realize it may be the point at which our masters press the reset button and start another game.

It is a horrible idea, and probably completely wrong. But the fact that there is even a smidgen of a possibility that we might be living in a Matrix universe is enough to cause a long dark night of the soul, should you start thinking about it too seriously. The only sensible advice is: don't. And if you do, hope that, as Zaphod discovers, the whole thing was created for your personal enjoyment, or at least survival.

12

parallel worlds

The Hitchhiker's Guide to the Galaxy has, in what we laughingly call the past, had a great deal to say on the subject of parallel universes. Very little of this is, however, at all comprehensible to anyone below the level of Advanced God, and since it is now well-established that all known gods came into existence a good three-millionths of a second after the universe began rather than, as they usually claimed, the previous week, they already have a great deal of explaining to do as it is, and are therefore not available for comment on matters of deep physics at this time.

Mostly Harmless

Go to the Equator, or at least to somewhere in the tropics. Failing that, find somewhere very, very far from any town or city or railway or road. Somewhere cold will do, provided the sky is clear. Wait till it is night. Now look up. You see the night sky in all its majesty. The great silver band that is the Milky Way, our galaxy's central disc, spread out over the sky. Perhaps 10,000 stars visible individually with the naked eye, and countless trillions reduced to smudges of light by distance and intervening dust and gas. It is an awesome sight, and the more one learns what one is actually looking at, the more awesome it becomes.

A few centuries ago, people thought those pinpricks of light were holes in the firmament, windows to heaven a few tens of kilometres above our heads. How dull! How small! Now we know they are stellar furnaces and galaxies, billions and trillions of kilometres distant. Now imagine for a moment that the sheer wondrousness that you can see in any African sky or through any telescope is no more a glimpse of the true magnitude of All There Is than a brief look down the road at the house next door.

We may live in a cosmos which is not only vaster than we can possibly imagine, but which is actually far vaster again – a Universe so big, so terribly huge and so complex that it encompasses many trillions – perhaps an infinite number – of alternate realities. Welcome to the possibility – probability – of the parallel universe.

That parallel universes may be real and not some silly sci-fi plot device (or a smoke-and-mirrors philosophical trick) is one of the most surprising things to have emerged in the past hundred years. To grasp the hows and whys of this idea involves some of the most difficult mathematics on the planet, and, as much of this is based upon quantum theory, necessarily it is stuff that no quite normal person can get to grips with. Nevertheless, the basic idea is quite simple, and has permeated the public consciousness to a surprising degree. Indeed, the idea of a shadow world is so firmly entrenched in folklore, religion, legend and entertaining fiction that the announcement by physicists and mathematicians that parallel universes may be real is akin to scientists at CERN or Brookhaven announcing that they had managed to use their particle accelerators to determine the number of angels mustered on the head of a pin.

Britain's Astronomer Royal, Sir Martin Rees (who has been much exercised by the notion of alternate realities), puts it thus:

> Some might regard other universes – regions of space and time that we cannot observe (perhaps even in principle) as being in the province of metaphysics rather than physics. But I argue that the question 'Do other Universes exist?' is a genuine scientific one.

Parallel universes are put to much use in the *Hitchhiker's Guide* series. Poor Arthur Dent, for instance, finds to his cost that coming from Galactic Sector ZZ9 Plural Z Alpha means that all sorts of unfortunate things happen to him. He loses the planet he loves to the Vogons and the girl he loves to the endless bifurcations of the Universe into an infinite stream of parallel worlds. As Vann Harl, the unpleasant new Editor-in-Chief of the Guide says to Ford Prefect, 'You've got to learn to think multi-dimensionally'.

To think multi-dimensionally we first need to deal with some semantics: the word 'universe' means 'everything there is'. If there happen to be other 'realities' – parallel worlds emanating in other big bangs or in other dimensions from ours, then they are, strictly speaking, part of our Universe. In the *Guide* the 'Whole Sort of General Mish Mash' gets across well many physicists' suspicion that outside, outwhere and outwith the realm of space, stars and galaxies that we see is a far vaster entity than we can easily comprehend (as if the 27 billion light year magnificence of the visible cosmos were not incomprehensible enough). In which case, we are looking for new terms to describe the entity that we have hitherto called 'our Universe'; perhaps 'metagalaxy', as Rees suggests, will do. (When the existence of galaxies beyond our Milky Way was first shown in 1920, the old term 'island universes' was initially used to describe these smaller units. That is part of the trouble with discovering endless new things – you have to keep finding new words for all the old stuff.) But, as Rees points out, until the existence of a larger and grander domain can be proved, we may as well leave our good old Universe be – and instead find a new word to describe everything else plus the stuff we know about. To date, the term 'multiverse' (Rees's neologism meaning the whole multiplicity of universes) has passed into general acceptance.

Many, if not most, societies throughout history have had myths about a world just around the corner from the real, tangible one that we live in. The shamanistic, animist religions that seem to be almost universal in pre-technological societies tend to focus on a spirit world, a place where the ghosts of ancestors mingle with the spirits of animals, plants, rocks, wind and rain. It is this idea of another world, hidden from ours, inaccessible except through prayer, trance, magic or in some cases the use of natural hallucinogens, that seems to be common to all religions – more so than a belief in a deity as such, or even an afterlife. (An oft-quoted but perhaps wrong idea is that humans invented religion when the awful reality of our mortality first struck us.) In more formal religions, the concept of another world becomes if anything more entrenched. The Norse gods lived in Valhalla, which was located at, but perhaps not really in, the mountains of Norway. The Greek pantheon lived, of course, upon Olympus. Did the Ancient Greeks actually believe that if you climbed Olympus (not, apparently, a particularly difficult task, even without fleeces and expensive high-tech walking boots) you would actually run into Zeus, Athena, Apollo and the rest having a chinwag and meddling in the fates of men the world over? Probably not. They stuck them on Olympus because it was high and conveniently out of the way and no one would think too hard about going to actually find them because of course if you go looking too hard for deities, like fairies at the bottom of the garden, you probably find ... er, diddly-squat.

It would be a mistake, though, to believe that people no longer take this sort of thing seriously. In Iceland, for instance, there still seems to be a fervent belief in a sort of strange, meta-Iceland, superimposed on the real one (as if the real one wasn't strange enough), populated, naturally, by elves. That said, one must always be careful when relating modern Icelandic folklore, as much of it tends to be imparted in modern Icelandic bars after the consumption of large quantities of frighteningly expensive alcohol, but there does seem to be a general consensus concerning elves.

Fiction writers love parallel universes; they can work as tremendous plot devices. What better than an entire new world, with its own morality, physics, love and everything else, which can be entered as simply as catching a train from an odd platform at King's Cross? Perhaps the best known crop up in children's literature, places where protagonists escape the ordinariness of adult dictatorship for brief bouts of unfettered freedom. Think of the land of Narnia, accessed, oddly, through a wardrobe. Narnia is full of baroque castles, exotic wizards and sexy queens, and it appears to run on a different timeline from the real world. When the children emerge after spending many years in Narnia, holding royal office no less, they find that no time at all has passed on Earth.

Ms Enid Blyton gets a bit of a knocking for her priggish morality, political incorrectness and simplistic language, but to my mind the best of her stories concern *The Faraway Tree*. This big old tree in the wood is, among other things, hollow and home to dwarves, and underneath is some sort of jewel mine. At the top is that most marvellous of things, a portal into alternative universes. Like some kind of enormous merry-go-round, a series of strange worlds whisk past the ladder extended from the topmost branch and transport Moonface (a jolly saucepan-selling chap) and his friends into a bizarre world where things are oh-so-different. Malarkey, of course, ensues; death (or at least mild unpleasantness) is defied; and everyone gets home safe in time for tea.

Perhaps the most famous parallel universe of all is that dreamt up by the Oxford mathematician Charles Dodgson in 1865. Via a wormhole-like underground portal a young girl enters a strange universe where the laws of physics differ substantially – allowing such entities as intelligent, waistcoat-wearing talking rabbits, mentalist milliners and disappearing cats. Dodgson, better known by his pen-name Lewis Carroll, was of course unaware of the revolution that was to occur in just 40 years' time thanks to the work of Einstein, but he knew enough geometry to understand the principle of 'multiply connected spaces'. Happily for Alice, her 'wormhole' stays open long enough not to crush her to a singularity.

In the *Hitchhiker's* books, Earth is destroyed by the Vogons, and a replacement planet is called into being by the Magratheans, on the orders of the mice. Later on, we learn that Arthur ends up on Earth at a date *after* the apocalypse. Is this the new Earth, commissioned by the mice? No. After the mice realize that the Ultimate Answer and Ultimate Question cannot coexist in the same universe they pull the plug on the whole project anyway and go back to philosophy, much to everyone's relief, especially the philosophers. Earth lies in ZZ9 Plural Z Alpha, a dangerously unstable region of space–time where anything can happen. Our planet is a world with many parallels, as Arthur discovers when he embarks on a long quest to find his vanished life (and vanished girlfriend Fenchurch), travelling repeatedly to the same coordinates, only to find a place that is mostly, but not wholly, unlike the Earth. Usually the result is unpleasant, perhaps no more so than when he arrives on a dull, soggy version of our planet called NowWhat and manages to get bitten on the leg by Agrajag, masquerading as a boghog. (Agrajag is someone else who probably wishes the Universe would just sort itself out and decide exactly who and what it wants to be and just get a life.) Later on, the infinite nastinesses of parallel worlds kick in again when Arthur is confronted by a daughter he never knew, a daughter who then goes on to discover that in the strange world of the plural zones one can discover two mothers – one who *did* go off with Zaphod at the party and then go on to become something of a hotshot in galactic media circles, and one who did not, tried to make it in New York instead, and for whom life turned out very differently.

So far, all of this – the Icelandic elves, Valhalla, the disappearing Fenchurch and the rest – has all been strictly in the realm of fiction, but can there really be any scientific basis for thinking that, just round the corner of the nearest molecule, there is an entire universe where everything is almost, but perhaps not exactly, just the same as in this one?

Yes. Parallel universes may be possible and necessary. Talk of them no longer gets you thrown out of the Physics Society bar: it gets you bought a round.

Broadly speaking, there seem to be four basic kinds of parallel universe: those that are a long way away; those that share our basic 'Mish Mash' but which are isolated by a bubble of different space–time; those created by quantum strangeness (and therefore lie just around the corner); and those that are just so peculiar that you need a brain the size of a small supercluster to understand how and why they are there. The simplest kind works on the principle that space is infinite – or at least something very like infinite (and things that are like infinity turn out to have remarkably similar properties to the Big One itself). How do you like the fact that there may be a million other Earths out there, each one containing you, or someone highly like you, living your life and having your experiences? In this scenario, you do not need to invoke anything truly bizarre: no parallel dimensions, no lurking behind a molecule or alternative timelines; all you need to assume is that space is extremely big. Of course space is big. The most distant objects we can see are about four trillion trillion kilometres away. There are more than a quadrillion galaxies within that enormous volume. That is a lot of stuff, and a lot of places for things to happen. That distance is limited by the power of our telescopes, but remember that there is a theoretical limit to how far we can see, even with the largest and best telescope imaginable. That limit is defined by the distance light could have travelled since the Big Bang. In effect, this defines a spherical shell centred upon the Earth beyond which objects appear to be disappearing faster than the speed of light – and which are hence unobservable (since we could never receive a signal from them). This shell is called the Hubble Volume; it is 27 billion light years across.

The size of this Hubble Volume is not fixed. If the expansion of the Universe continues, at a rapid but slower-than-light-speed rate, more and more galaxies will come into view. The observable Universe will get bigger over time. But if the rate of expansion is accelerating – as seems to be the case, thanks to dark energy – then over time the most distant galaxies will be literally whisked away from view. Our visible world will in effect appear to contract, even as the Universe gets larger.

There are unlikely to be parallel Earths within our Hubble Volume. Let's assume that it would not be crazy to suggest that there are as few as, say, ten active advanced civilizations in each large galaxy such as ours. The chances of any of these Milky Way species being anything like us are, of course, extremely small. It is not impossible that there is a planet identical to Earth within our galaxy, but we can discount this idea on probability grounds. But of course there is more than one galaxy, which means that there are probably trillions of civilizations out there. Estimates range from 10^{20} (a *Star Trek* universe) to, er, seven. But what are the chances of any of them actually being identical to ours? Probably zero. And even if there were a planet very like ours out there, it could not be identical; it would be sitting in a different part of the cosmos, having had a quite different timeline. In *Star Trek*, there are lots of humanoid creatures out there, which differ from us only in that they have funny foreheads and strange accents. In reality, even this level of similarity is extremely unlikely.

So we – and the world around us – are unique. But only in the part of the Universe that we can see. If space is essentially infinite – with an infinite number of stars and galaxies to go with it – then at a distance about 100 times further than the visible horizon of our Universe there should, statistically, be at least one 'Earth' identical in all respects to this one, including a 'you' reading this book right 'now'.

This is because when you have infinity to play around with, anything is possible – and indeed certain. There will be an infinite number of 'yous' out there – not just creatures that look like you and live on planets that are quite like the Earth, but beings who look exactly like you and have had exactly the same life as you, with exactly the same thoughts, feelings and emotions in exactly the same order. Furthermore, the skies they will gaze out from on their pseudo-Earth will be identical in every respect to the skies that you see. Given enough space and matter, you will eventually get to a point far beyond (but not impossibly far beyond) our visible horizon from which an *identical* Hubble Volume – with the exact-same planets, stars and galaxies – appears to surround you. You do not even need an infinite cosmos for that – just one say 84 orders of magnitude larger than the one we think we live in. So, this theory says, you should have an identical twin, living in an identical universe, to all intents and purposes, some 10^{115} kilometres away. (This number is derived from a calculation that uses the total number of protons that could be contained within a Hubble volume, and assuming that each of these protons may or may not be present in any given volume, so all possible permutations are accounted for. This calculation does not take into account quantum states, which may be infinitely variable and which some say influence, or even constitute, that tricky business we call consciousness. If quantum states are intrinsic to thoughts and experiences, your identical twin will live even further away.)

We can call this the brute-force scenario. It involves nothing very clever: no quantum strangenesses, no hidden dimensions, no mysterious eddies in the space–time continuum. All it requires us to imagine is that the Universe is bigger than the relatively small volume that we can see, that matter is distributed within it, and that there are no meaningful structures larger than about a quadrillion quadrillion yards across – i.e. on a large enough scale, the Universe is basically boring and homogeneous. Unlike all the other parallel universe scenarios, this twin-you is, in principle, observable. All we have to do is wait a few trillion years for the nearest Hubble Volume identical to ours to pop into sight – that is, if dark energy stops or slows down for some reason. If it does not, then most of what we see even now will zoom over the horizon, and when the Universe is roughly double its current age, the only visible objects will be the stars and galaxies in our local supercluster.

It is at this point that it is worth explaining why we should want to believe in parallel universes. There seems no compelling reason to imagine that there is anything out there beyond what we can see and detect with our instruments. Reality, with its laws and complexities, is impressive and big enough as it is, surely, without us having to imagine other versions of it. And, don't parallel universes simply fail the test of Occam's razor, that wonderful tool that pares down entities to the minimum possible? If we are going to dismiss God on the basis that using Him as an explanation for why and how we are here is poor logic, on the grounds that He now needs explaining, then surely we are going to lead ourselves down the same, problematic garden path with parallel universes?

In fact, there is a good philosophical, metaphysical and even plain old physical reason to invoke parallel realities. The anthropic principle (see also Chapter 4 on the existence, or otherwise, of God) neatly encapsulates why we might need reality to be more than it seems. In short, the Universe looks suspiciously as though it is 'biophilic', finely tuned to allow the emergence and existence of life.

So why is this a problem? Surely the answer is simply that 'We are here, we are thinking about this, so we just happen to be living in a universe which evolved in such a way as to allow life to exist. If it hadn't we wouldn't be here, and there would be no problem. QED'. Fine, except physicists don't relish the 'sheer, random coincidence' solution to the anthropic problem. Say, for example, you are wondering about the size and mass of the Sun. You realize that if the Sun were just a little bit bigger – say twice as big – it would be so hot that life on Earth would be impossible. And if it were just a bit smaller – say two-thirds the size – then Earth would be colder than Mars is in our reality. And we know that it is very easy for suns to be much, much bigger than ours, or much, much smaller. The smallest viable stars are some ten times smaller than the Sun, and the biggest a hundred times larger. But we are lucky enough to have a 'Goldilocks' Sun – neither too hot nor too cold. Should we be suspicious?

Of course not. The sky is full of stars, some big, some small, some just the right size. It is not bizarre at all that we happen to live on a planet that orbits just the right kind of star. If we didn't, we wouldn't be here (or would have evolved into a creature which could tolerate Venusian or Martian conditions). We know there are plenty of stars where planets orbiting at 150 million kilometres *would* have the climate of Venus, or of Mars. There is no anthropic problem here because all the other alternatives are not only possible but real.

But having a Goldilocks universe *is* a problem, just as it would be if there were only one Sun. If you only have one Universe and that is perfectly tuned for life, the most obvious solution (apart from horrible coincidence) is God, and most scientists don't go a bundle on that idea either. If you are happy to imagine hundreds of different universes, many of which do *not* contain the right conditions for life, then the anthropic problem disappears.

A true multiverse consists of universes so separate that different physical laws and constants prevail, thus allowing a large number of initial conditions, laws of physics and so on from which our biophilic universe could 'choose'. This of course assumes that other physical laws and constants are possible. This may not be so. Einstein's assistant Ernst Strauss once asked, 'Did God have any choice when he created the Universe?' – a very good question indeed. If the answer is 'yes', then the anthropic problem can only really be solved with a multiverse. If it is 'no', then the problem disappears. It may be that what we call the 'laws of physics' are not absolute but are merely what Rees has termed 'bylaws'; local interpretations of some fundamental rulebook which we have not yet deciphered. If this is the case then we may need to invoke the strangest kind of multiverse of all, one anchored only in the bedrock of mathematics and logic rather than in observable reality.

One controversial but attractive idea is the 'many worlds' interpretation of quantum mechanics, which we met in connection with time travel in Chapter 7. Unlike all the other parallel universe scenarios, this one is truly odd. To get to your identical twin in the brute force multiverse, you would simply need a spaceship built to last and a very long time. To get to a quantum parallel universe you would need to travel no distance at all, yet these worlds will remain forever beyond reach.

The idea is an attempt to explain the manifestations of the quantum world in real terms rather than purely mathematically. Quantum events are things like the emission of a photon when an electron changes its energy level, and it seems that the outcome of these events is far removed from the neat way in which larger objects behave. Broadly speaking, quantum theory states that the particular state of an event or object may be best described as a 'wave' of probability; at any one time, the actual position of, say, an electron is not only unknowable but actually indeterminate. The act of observation itself – be it by a human, a machine, or by another particle – causes, in one interpretation, the wave to 'collapse' into a hard-and-fast here-and-now location. Hence the famous paradox of Schrödinger's cat. Put a cat (or any other animal) in a sealed box. Inside the box is a loaded gun (or some other lethal device, such as a cyanide pellet). Next to the gun is a machine designed to pull the trigger when a quantum event, say the emission of a particle by a radionuclide sample in a given time occurs. Now, since events like this are non-deterministic – alpha particles, say, are not fired off because something 'happens' in the lump of radium to make this occur – we cannot say at a particular time whether the cat in the box is alive or dead. Only when the box is opened will the wave function 'collapse', and the outcome of the non-deterministic quantum process be revealed.

It could mean that as soon as you close the box, the gun either fires, or it does not. You have a dead cat, or a live one, and you find out which when you open it a minute (or whenever) later. But according to this interpretation of quantum theory, the fact that the inside of the cat box is truly isolated from the rest of the Universe (and ignoring the fact that to do this one would need to put it in orbit around a black hole the far side of Pluto, or something equally implausible), means not just that do we not know what state the cat is in, but that it is actually not in any particular state until we look inside (ironically, the cat experiment was devised by Schrödinger to show just how ridiculous the idea of quantum superposition was). There are, in other words, two cats inside the box – indeed, two boxes – one alive and one dead, something called a superposition. The quantum uncertainty created by a single yes-or-no event has, in one interpretation, produced two whole universes. In the other universe – manifested at the moment the gun was either triggered or was not – the experimenters open up the box and find a dead cat where you have found a live one – or vice versa.

This sounds silly, and indeed spooky. Why should the act of observing alter reality? If it is the case that on the quantum level objects like electrons can be anywhere and everywhere, then what does this really mean? What about bigger objects:

your car or kitchen table? They are made of quantum stuff, yet when you get back to the car park after doing the shopping you never find that your Volkswagen has mysteriously relocated itself to Neptune (unless you live in some of the dodgier bits of North London).

In essence, the many worlds interpretation of quantum theory states that when a particular event can have several possible outcomes, *all* outcomes happen every time. The weirdness disappears because cats are not dead and alive and tables are not in two places at once, it is just that new universes are ceaselessly being created in which all possible quantum states occur. This differs from the interpretation that says that there is only one real outcome, the one which the wave function collapses into. This many universe idea, first proposed by Princeton physicist (actually only a student then) Hugh Everett III in 1957, has profound implications. The number of parallel universes in existence would effectively be infinite. Every single 'choice' made by every single particle in the Universe would have been made. In your universe you come out of the car park and find your trusty steed still there. In a (much smaller) number of universes you find it is not, curse and never discover that a rather bewildered automobile has suddenly materialized in the outer wastes of the Solar System. On the macro-scale, this would mean that every possible 'version' of reality is equally valid and has actually happened, and is continuing to happen. There was thus a universe where the Soviets did not turn the ships around and Krushchev and Kennedy went to war. There was another one where commonsense prevailed in Vietnam. There are countless universes where there never was a Cuban Missile Crisis, or a Vietnam War. An uncountable number of realities contain no organized matter at all.

Does this deal with the anthropic problem? Although we have created an infinite number of universes, it seems that the basic physical laws, constants and so on may have to be the same in every one of them. No quantum event can change the universal constants. We may just have created a very large version of our own Universe without really solving the problem at all. Nevertheless, it is an awesome thought.

Where are these other universes? In the quantum many worlds multiverse they jostle alongside our Universe in something called 'infinite-dimensional Hilbert space', a mathematical construct that houses the wave functions of quantum events. Travel from one parallel universe to another is impossible, but only in the sense by which we normally mean travel. From a godly perspective, there is only one true universe here – one with an infinite number of states.

Some object to the many worlds hypothesis. People say it is simply 'too crazy, too wasteful, too mind-blowing', David Deutsch, of Oxford University, told *New Scientist* in 2002. 'But this is an emotional, not a scientific reaction. Quantum theory leaves no doubt that other universes exist in exactly the same sense that the single Universe we see exists. This is not a matter of interpretation; it is a logical consequence of quantum theory.'

Confused? It gets worse. As we saw in Chapter 6, the inflationary event that gave rise to the Universe may have been on such a large scale that our Big Bang is a merely parochial, trivial event, There are an infinite number of bubbles, 'banglets' out there, so the presence of twin universes is assured. Yet these universes (even though they occupy the same basic space continuum as ours) are as inaccessible as the many worlds formed by quantum events. Even if you travelled at the speed of light in a spaceship forever, you would never reach one because the fabric – the 'dough' in which they are embedded – is expanding faster than the speed of light. Many of these bubbles may have quite different laws of physics. Hence the anthropic problem disappears.

There are other multiverse scenarios. Lee Smolin, Professor of Physics at Pennsylvania State University, has suggested that black holes may effectively create new universes within their event horizons. In M-theory, the 11-dimensional supercharged version of string theory, our Universe is just one of many caused by the collision of 'membranes' in 11-dimensional space. If so, the laws of physics may be arbitrary and unique to each individual universe. It may be that all possible universes, obeying all possible 'solutions' to basic mathematical laws, can and must exist. According to the Platonic paradigm, it is mathematics, not reality (or observed reality) that is fundamental. In the beginning there were the equations, and they were good. Everything – and anything – flowed from that. The ultimate reality is not a force, or a description of strings or branes, but mathematical structure existing wholly outside of space and time.

Does any of this matter? If these parallel realities are unvisitable and unobservable, they do, as Rees concedes, lie in the realm of metaphysics. But the true nature of our cosmos might yet be revealed by our ever more sensitive instruments, instruments like WMAP, COBE and Hubble. A few decades ago, Isaac Asimov wrote a brilliant science fiction story, *The Gods Themselves*, in which a parallel Universe is discovered and differences in the physical constants used to create a source of energy that mutually benefits intelligent beings in both realities. Maybe it will be possible after all to move from one parallel realm to another. Which would be a more profound shock to humanity – the discovery of alien life in this Universe or the discovery of an alternate Earth where Tricia McMillan never goes back to get her handbag and so misses her opportunity to run off with Zaphod Beeblebrox?

13

the whale that came from nowhere

And wow! Hey! What's this thing suddenly coming towards me very fast? Very very fast. So big and flat and round it needs a big, wide-sounding name like ... ow ... ound ... round ... ground! That's it! That's a good name – ground.

I wonder if it will be friends with me?

Last thoughts of the unnamed whale called into existence by the Infinite Improbability Drive

It is little appreciated in our seemingly orderly Universe that just about anything can happen at any time. It is quite within the rules of mathematics, if not of casinos, for you to be dealt a perfect hand of cards, or to bet on red 4,000 times in a row and walk out of the casino with more dollars than there are atoms in the known Universe. Of course, if you actually did this you would be taken down into the casino car park and kneecapped by large men in shiny suits. It might just be that the first bunch of monkeys you round up and equip with typewriters do indeed type *Hamlet* first time, word-perfect.

You can win the lottery 13 times in a row and not one law of physics will be broken. All the atoms in the car you are driving may suddenly relocate themselves one metre to the left, leaving you with an embarrassed look as you roll to a painful and probably fatal halt on the motorway. You may wake up on Jupiter, or inside the stomach of a whale itself called into existence only a few seconds previously. Surpris-

ingly enough, thanks to the rules of probability (and especially – when it comes to things like waking up on Jupiter, the
oddities of quantum physics), all of these fantastically
unlikely things are not actually banned.

Not long ago a headline appeared in the newspapers along
the lines of 'Big Bang in your cornflakes'. The story was based
on a lighthearted calculation in one of the scientific journals of
the odds of an entire universe popping into existence just as
you sit down for breakfast. The odds, while small, are not infinitely small (we are talking the reciprocal of ten to the power of
a number commensurate with the total number of particles in
the Universe, then again to the power of itself). And very
strange things do happen, every day. It may be rare for whales
to appear out of nowhere, but it is extremely common for electrons to be in two places simultaneously and an extremely
improbable quantum event may even have given rise to our
entire Universe.

The Odd Ship *Heart of Gold* managed to turn the gigantic
numbers associated with improbable events – such as the Universe's sole surviving Earthwoman rescuing the sole surviving
Earthman just as he has a second left to live in the vast, incomprehensible emptiness of interstellar space at an improbability
level measured as two to the power of the seven-digit telephone number of the flat in Islington where they had first met
– into a motive power source of unparalleled vigour.

Adams's satirical use of probability as a means of propulsion is
extremely apposite. Today, more perhaps than at any time in
our history, an understanding of probability is necessary to
make ethical, economic and personal judgments. In the old
days, we had only a few probability equations to deal with.

There was death, which was certain (still is). It has a probability of one, or 100 per cent, depending on how you talk about these things. Ditto taxes. And many other things central to life were certain, and comforting in their certainty – the passage of the seasons, the reproductive cycles of crops and animals. In the pre-technological age – and in much of the world still today – probability could be safely ignored, in favour of fatalism toward risk and danger. If you have no access to medicine, clean water and nutritious food, you have no real control over the hazards you face every day. It is likely that some of your babies will die. It is probable that during your life you will suffer a number of quite unpleasant diseases and that sooner or later there will be one from which you will not recover.

These days we talk not of dangers or the unavoidable and messy biological realities of existence, but of risks. And having evolved on the African savannah we find ourselves spectacularly incapable of making even the most inaccurate stab at assessing the complex, and often hidden, risks of the modern world. *Homo modernus*, in fact, is a simpleton when it comes to gauging the probability of anything except the chances of the Sun rising each morning.

Every now and then someone does a survey attempting to quantify the public's evaluation and prioritizing of risk. Each time, the results make depressingly familiar reading. Asked, for example, to rank a number of health dangers, a large group of American students in the early 1990s came up with a telling result. At the top of the list were things like 'living next to a nuclear power station', and 'pesticides'. Moving down the list we have things like 'radiation from computer monitors' and 'food additives'. Only then do we get to some significant hazards like riding a bicycle and driving a car. At the bottom lurks stuff like 'smoking', 'swimming' and 'home improvement'.

Nearly all these surveys show that we get risk rankings back to front. The things society worries the most about – nuclear power, air travel, pollution, train accidents – are statistically unlikely to harm us. We worry far more about plane crashes (a given flight on a large commercial airliner stands about a one in a few-million chance of killing us) than the drive to the airport (the risk is a few tens of thousands to one against). We worry far more that our children will be killed by strangers walking on the streets alone (the chances of this happening are about one million to one against per child per year) than that they could be run over by a parent driving their own poppet to school for fear that they be exposed to the dangerous streets – the chances of a child dying in a motor accident are at least 100 times greater than of being murdered by a stranger. And when children *are* murdered, they are about half a dozen times more likely to be killed by their parents or a close relative than by someone they do not know. We worry more about pesticides than smoking, more about being mugged on holiday than falling drunk into the hotel pool. We fret far more about food additives than about getting a healthy balanced diet.

Probability does some really weird and counter-intuitive things when really large disasters are involved. For instance, asteroid strikes on the Earth are popularly considered to be just a cheesy sci-fi plot device. In Britain, Lembit Opik, an MP who has campaigned vigorously for some sort of government-funded early-warning system to be put in place, is commonly dismissed as a space cadet. Yet Opik is quite right to be worried. The point about big asteroid hits is that they are both extremely rare (reassuring) and terribly catastrophic when they do happen (less so). A strike of the magnitude that is thought by some to have seen off the dinosaurs 65 million years ago probably only happens every 100 million years or so. We have certainly not experienced one since, and may not do so during the remaining sojourn of *Homo sapiens* on this planet. But if one were to occur tomorrow then we can assume that maybe five billion people would be killed. If you do the sums, it turns out that for any person alive today the chances of kicking the bucket in an asteroid strike are 750 times higher than winning the British National Lottery or around twice as likely as being killed in a plane crash. Yet we spend millions on lottery tickets, billions on aircraft safety and thruppence on trying to spot asteroids.

The issues here seem to be *control* and *frequency* (as distinct from probability). We are more frightened of planes than cars because we drive cars and don't fly planes. Something about having someone else at the wheel makes us feel more vulnerable – many people find that even a couple of lessons dramatically reduces any fear of flying. Ignorance, far from being bliss, breeds fear; knowledge sows reassurance. We don't worry about asteroids because no one has ever seen an asteroid strike. (Actually that is not true; something very nasty, maybe a cometary fragment, was witnessed slamming into the atmosphere over the Tunguska region of Siberia in 1908; if it had hit eight hours later it would have flattened Edwardian London, which is on the same latitude.) Ditto pandemics (hardly anyone is still alive who remembers the 1918 influenza disaster).

The media play a major part. Journalists give accidents in atomic power stations much more prominence than those in coal or gas plants, even if the mishaps have nothing to do with nuclear power *per se*. In late 2004, a nasty accident in a Japanese atomic plant killed a handful of workers. The item received top billing in the news, despite the fact that the cause of the accident – a steam escape – was not related to problems in the reactor core, and there was no release of radiation. The cumulative effect of this media-driven hype of accidents caused by the unusual, glamorous or exotic is to push up the perceived risk of some technologies and activities (air travel, railways, atomic power, genetically modified foods, vaccines) at the expense of others (power tools, driving, excessive alcohol consumption, rugby).

The result is that much of our public policy and, in particular, decisions about how we spend our tax pounds/euros/dollars these days, particularly in areas like health and transport, is predicated on a total misreading – often wilful – of statistics. Probably the best example is railway safety. We all know, deep down, that trains are very safe, yet we demand a far higher threshold of safety from them than we do of, say, cars, bicycles and motorbikes. After every fatal train crash in Britain, there are calls for hundreds of millions of pounds of taxpayers' money to be spent to prevent a reoccurrence. Little heed is given to the fact that by demanding almost 100% safety from public transport, we need to spend so much money on it that some must be recouped in higher fares, in turn forcing many people out of the trains and buses and on to the roads – where the chances of them dying are far, far greater. Statisticians talk about cost per life saved. On trains, for instance, that figure is often a hundred or a thousand times higher than for road travel.

At the same time, probability – chance or hazard, its old names, gave more of a feel for the slippery nature of the beast – is proving to be a powerful tool. Astronomers use probability to analyse the patterns seen when large telescopes, or arrays of smaller telescopes, build up huge portraits of the sky thousands of light years deep. Probability is used by organic chemists looking to synthesize new drugs and by anti-terrorism experts wanting to discern patterns in over-the-counter pharmaceutical purchases in American cities that might show that a bioterrorism attack is under way.

Data mining, a form of statistical analysis that requires the computerized (in practice) number-crunching of vast data sets to discern otherwise hidden patterns – in people's credit card purchases, airline bookings, car rentals and other often quite personal aspects of the populace's behaviour – has been recognized as such a powerful and potentially malevolent tool that some aspects of it have been specifically banned by US law. Similarly, in a 1996 rape trial an English judge attempted to prevent Bayes' Theorem being used as evidence when considering DNA testing and the probabilities of the suspect being guilty.

Probability is hard to get to grips with. Most of us get no further than the mathematics of the card game or roulette wheel, leaving actuaries and bookies to make a killing on the difference between perception and reality. Adams got it right. Probability is indeed a powerful and, in the right hands, sharp tool, separating the relevant from the irrelevant, the dangerous from the innocuous, and the likely from the absurd; it can save your life and make you a fortune. In the wrong hands, probability, as Zaphod and Trillian found, will take you to some very strange and dangerous places.

14

ultimate questions – and answers

It had to be a number, an ordinary, smallish number, and I chose that one. Binary representations, base thirteen, Tibetan monks are all complete nonsense. I sat at my desk, stared into the garden and thought '42 will do'. I typed it out. End of story.

Douglas Adams

Type 'answer to life, the universe and everything' into the Google search engine and its calculator function is prompted to respond '42'. This number is of course the answer Deep Thought gives after pondering its navel for a rather inconvenient seven and a half million years. The mice (who built the computer) are not happy, as Deep Thought had warned them. 'You aren't going to like it', he says as his monitors come on for the first time in half a million generations.

Boiling down Life, the Universe and Everything to a single number is a good joke. Adams was playing with the idea that everything physics knows and wants to know about the cosmos can be summed up by a few constants. In 1999, Sir Martin Rees wrote a book called *Just Six Numbers* in which he examines the key quantities that shape our cosmos (42 is not one of them, sadly). The Ultimate Answer must surely be about more than just a bunch of numbers? To know what it is we will, as Deep Thought realized, have to work out what the Ultimate Question is. Here at least we do appear to be making a little progress.

Are the laws of physics that we observe all the laws that there are? Is there an as-yet undiscovered set of axioms underlying the laws as we observe them? Are other laws possible or is this Universe the only type of universe that there can be? Physicists talk about the 'symmetries' of nature, many of which have been lost since the Big Bang – there is no reason to suppose, for instance, that equal quantities of matter and antimatter were created at the beginning. There is the eternal mystery of time. What exactly is it? In *The Time Machine*, H. G. Wells articulated the view that time can be treated just as another dimension, like the three spatial ones we are familiar with. But our experience of time is not really as just another up, down, sideways or backwards. To us it has an arrow: it moves in only one direction. As Brian Greene puts it:

Eggs break, but they don't unbreak; candles melt, but they don't unmelt; memories are of the past, never of the future; people age, but they don't un-age. These asymmetries govern our lives; the distinction between forward and backward in time is a prevailing element of our experiential reality. If forward and backward in time exhibited the same symmetry we witness between left and right ... the world would be unrecognizable.

Perhaps, as Greene and others have asserted, time's arrow was set in motion at the birth of the Universe.

Then there is dark energy. More than two-thirds of the observed Universe consist of this, driving the galaxies apart ever faster. Is dark energy the same force that gave rise to cosmic inflation at the Big Bang? We know that the vacuum bubbles and seethes with fantastic energies – a natural candidate for this mysterious anti-gravity force – yet the strength of dark energy we observe seems to be somewhat weaker – by a factor of 10^{120} – than the equations predicting the size of the vacuum energy suggest. Is dark energy related to the Higgs field, one of the Holy Grails of modern physics? The Higgs field is supposed to give particles the property of mass. (The nature of mass is one of the big and embarrassing things, like time, that physicists still really don't have a clue about.)

And what about dark matter? Ever since it was realized that the shape and size of the observed galaxies could not be accounted for by the visible stuff – stars, planets and dust – the search has been on for this odd substance, which is thought to pervade the cosmos and to interact with ordinary matter only through the weak agency of gravity. Dark matter was discovered by the Swiss astronomer Fritz Zwicky, then working at Caltech. In the early 1930s he realized that a cluster of galaxies he was observing were not moving correctly. Rather than flying apart, as they should, they seemed to be bound together by an invisible force, presumably gravity, generated by some invisible substance wrapped round the galaxies like a halo. Little known outside cosmological circles, Zwicky deserves credit for increasing the scope of the Universe by an order of magnitude at a stroke.

Physicists would like to know whether there are more dimensions than the three of space and one of time that we see, or think we see. String theory – and its most up-to-date incarnations, superstring theory and M-theory – strongly suggests that there are another seven dimensions, probably rolled up so slightly that we have not yet noticed them. String theory says that instead of being made of point-like fundamental particles, matter actually comprises tiny filaments of pure space–time, several quadrillion times smaller than a proton, each of which is shaped and behaves like a little string. The Standard Model of particle physics says there are about 50 fundamental particles – a zoo, to be honest – including 12 species of fermions (matter stuff) and several bosons (radiation) plus their corresponding antiparticles, together with particles yet to be found such as the matter-carrying Higgs boson and the graviton. But the strange menagerie has a far from fundamental feel about it – hence string theory.

It is the ways in which these strings vibrate, the theory goes, that give all the different particles their properties – the charge of an electron, the mass of a neutrino and so on. If superstring theory is right, we have answered one of the key and most fundamental questions of science, namely 'What is stuff made of?' The only problem is that to probe matter on these tiny scales requires energies far in excess of what even our most powerful particle accelerators can manage – a true challenge for the 21st century.

A complementary Big Question is whether it is possible to unify the fundamental forces and particles into one. The Grand Unified Theory (or GUT) would end several decades of rationalization under which previously separate entities – such as the weak force and the electromagnetic force – have been shown to be merely different aspects of the same thing. Can we unify the physics of Einstein – of the very big, of gravity, space and time – with the quantum physics of Neils Bohr, Heisenberg and Planck, the Alice-in-Wonderland world where objects can be in two places at once and 'messages' can get from one side of the Universe to the other in an instant? To do so we will need a theory of quantum gravity.

What, exactly, *is* space, the entity in which all this sits? Isaac Newton thought that 'absolute space' was a fixed framework in which everything was located and against which all movement could be measured. In the 1870s the Austrian physicist Ernst Mach (of speed-of-sound fame) suggested that absolute space is an illusion; if you spin in the void, if you have absolutely nothing to be spinning relative to, then you will not feel any centrifugal forces. In his theory of General Relativity, Einstein showed that although Mach was on the right track, he was wrong about there being no reference frame without matter to define it. He introduced the concept of 'absolute space–time', a revival in a sense of Newton's original concept.

On the small scale, we know that quantum physics is right but we don't really understand what it means. What does it entail for electrons to be in two places at once? Is it really the case, as Richard Feynman said, that when you walk across a room you are trying out every single possible path from one side to the other, including those which take you via Pluto, and that the obvious path is just the most probable one? How does Einstein's spooky action at a distance actually work? If we live in an eleven-dimensional Universe, how exactly does this explain gravity, light and time? The idea that light – all electromagnetic radiation – may just be manifestations of ripples in the fifth dimension sounds as weird as mice being manifestations of creatures from a parallel universe, but it is taken seriously.

On one level, it is disheartening how many of these questions remain unanswered. On the other hand it is impressive that just a hundred years ago none of them had even been posed. Progress is certainly being made – at a price.

Probing the deep structure of the Universe is very expensive. To generate and examine the smallest, most fundamental particles, larger particles must be smashed together in some of the most costly machines ever built – particle accelerators and colliders. For decades CERN's marathon-length tunnel has housed something called the Large Electron–Positron (LEP) collider; under construction is an even more fearsome beast, using the same tunnel, called the Large Hadron Collider (LHC). When completed in 2007, this will be the most powerful particle accelerator in the world. Physicists hope to use it to ram protons together at an energy of 14 TeV (teraelectron volts) and answer many of the fundamental questions. It may unearth the elusive Higgs boson (aka the God Particle) and thus solve the mystery of mass. It may also determine whether neutrinos are massless and antimatter really is a perfect reflection of matter. It may even unravel some of the extra dimensions predicted by string theory (although even the energies of the LHC will not be enough to uncover the strings themselves). An even more powerful machine – the International Linear Collider – is at the planning stage, and when built will complement the LHC. Together, these two machines will be by far the largest and most awesome scientific instruments ever built, true 21st century marvels.

At the other end of the scale, probing the cosmos requires machines almost as large and expensive as the giant supercolliders. The probable demise of the Hubble Space Telescope in the next few years, arguably the most popular scientific instrument ever built because of the spectacular images that it has provided, leaves a hole that must be filled. Planck, a European spacecraft, is due to be launched in 2007 and will sit balanced by the Earth's and the Sun's gravities at a point called Lagrange 2, a couple of million kilometres from Earth. It will detect anisotropies – minute differences – in the cosmic microwave background radiation with far more sensitivity than WMAP. By the time the mission ends at the beginning of the next decade, we may know for sure what happened in those fleeting but all-important moments after the Big Bang.

One of the biggest questions of all of course is 'Are we alone?'. We may be able to answer this in the next couple of decades – but only if the answer is 'no'. The Search for Extraterrestrial Intelligence continues apace and thanks to improved computing power will probably find, by 2030 or so, anything within a couple of hundred light years from Earth that is actively transmitting radio signals. Of course a null result does not prove we are alone. It is a sobering thought that the human race could very well spend the rest of its existence in the dark about this. If alien civilizations are as rare as some scientists now believe, we have a very lonely future ahead of us. But even if civilizations are rare, life might be as common as muck. After 20 years in which Mars was written off as an orange version of our Moon, the bets are now off as to whether the Red Planet may be home to at least microbial life. Every two years, for the next couple of decades at least, NASA will launch new probes to Mars. By 2020, there should have been some 30 missions to prod, probe and drill its surface with ever-more-inventive robots. Ironically, discovery of life on Mars may well take place on Earth. Recent signs of methane in the Martian atmosphere – a key marker for biology – have been found using Earth-based telescopes. And of course there was meteorite ALH84001, which may or may not have contained fossil Martian bacteria. We had to go no further than Antarctica for that.

To my mind the most exciting future space mission (save a manned exploration of the Solar System) is the search for Earth-like planets orbiting other stars. Europe's planned Darwin mission will send a flotilla of six 150 cm space telescopes orbiting at Lagrange 2 (the same place as Planck will sit) to scan the nearby stars for blue–green worlds. Due to launch in 2015, Darwin has not yet received funding and the Europeans and NASA may end up collaborating on a joint mission to do the same. Even larger Earthbound telescopes should be able to analyze spectroscopically the atmospheres of small exoplanets. If they find chemicals like water, oxygen and methane then we will have a good candidate for life.

Not all the great mysteries are 'out there'. Perhaps the most perplexing scientific conundrum concerns the nature of human consciousness. What does it mean to think, to be self-aware? Some scientists feel that this is the biggest and most important question of all. Others will not even discuss it at the dinner table. Like all the big questions, it borders on philosophy and metaphysics. What is a conscious thought or experience? What are those enigmatic 'qualia' that give the neuroscientists such headaches? What does consciousness – indeed what does physics – say about free will? In the 1980s neuroscientist Benjamin Libet monitored brain and nerve impulses to show that the movement of muscles under conscious control – the tapping of a finger, for example – is directed by nervous impulses from the brain and spinal cord *before* we become aware of wanting to move. These results imply that our ideas about will and choice may be purely illusory; our consciousness may have no impact on our behaviour, but may instead be merely a passive observer. If we do work out what consciousness is, or even come up with a reasonable theory of how it emerges, will we ever be able to engineer it into a machine? A conscious computer or robot would, in many ways, be as staggering a development as the detection of alien civilizations (on a profound level it would amount to the same thing). What would a computer feel? Paranoia, like poor Marvin, always suffering that dreadful pain down his left side? Infinite superiority and smugness, like Deep Thought? Or aggressive hatred, like the machines in the *Terminator* movies?

What happens when we die? Here we get back to God again, but there are other solutions that do not involve either oblivion or the pleasant/unpleasant alternatives on offer from many of the major religions. In the 1980s, Frank Tipler (whom we met in Chapter 7) came up with a hypothesis called the Omega Point. At some time – the Omega Point – in the far distant future of the Universe, Tipler suggested, the cosmos itself will be harnessed by advanced civilizations (which by now have enveloped and controlled all the matter and energy in the Universe) as a giant supercomputer, and with its almost infinite computational powers it will be able to recreate every past event and object in its computational matrix. (There are similarities with more conventional matrix scenarios which suggest that computer modelling of entire universes could come a lot sooner than this.) This re-creation would include the past lives of every being – including us – that has ever lived, and hence we would have a computer-generated afterlife of infinite duration (and, he suggests, paradisiacal properties) to look forward to. It's a nice idea, if almost certainly crazy. Certainly something worthy of an entry in the *Guide*.

So, we still have a whole load of fundamental questions, the answer to none of which (as far as we know) is 42. The nature of matter, space and time. The origin and fate of the Universe. The laws of physics. The fundamental forces and particles. What it means to *feel* alive, and whether we are alone. As the multi-dimensional *Guide* tells a bemused Random in *Mostly Harmless*, 'All you really need to know for the moment is that the universe is a lot more complicated than you might think, even if you start from a position of thinking it's pretty damn complicated in the first place'.

A good candidate for the Ultimate Question seems, to me anyway, to be 'Why is there anything here at all?' The more you think about this question the deeper and more unsettling it becomes, and the ability to unsettle is usually a sign that you are on the right lines. For let's face it: when all the questions about the laws of physics and the fundamental forces and particles have been answered, when we have discovered the origin of the Universe and its fate, solved the riddles of time and space and met our extraterrestrial neighbours (if there are any), we will still be left with the elephant in the room, the big problem of why any of this at all should *be*. It is hard, but just about possible, to imagine another state of affairs, one where nothing ever was, nothing is and nothing ever will be. Being, or not being, that really is the question. We will probably have to think of something even cleverer than Deep Thought to find the answer to that one.

further reading

on the possibilities of alien life

Barrie W. Jones (2005) *Life in the Solar System and Beyond*. Springer-Praxis, New York.

Quite technical, but brings you right up to date on the current thinking as to the whereabouts – and probability of the existence of – ET. Everything from how to find exoplanets to the chances of life in our Solar System, to what to do should we get a message from alien intelligences and the likelihood of this happening anytime soon.

Peter Ward and Don Brownlee (2000) *Rare Earth: Why Complex Life is Uncommon in the Universe*. Springer-Verlag, New York.

Why intelligent life may be more unusual than we thought. A convincing solution to the Fermi Paradox which points out just how special our planet is.

Stephen Webb (2002) *If the Universe Is Teeming with Aliens – Where Is Everybody?: Fifty Solutions to Fermi's Paradox and the Problem of Extraterrestrial Life*. Springer-Verlag, New York.

Webb presents his bewitching thesis that we might be alone, after all. Counter-intuitive and fun.

Michael Hanlon (2004) *The Real Mars*. Constable, London.

A full account of the Red Planet, including the probability of life on the fourth rock from the Sun.

Sir Martin Rees (2003) *Our Cosmic Habitat*. Phoenix, London.

A brilliant run-through from Britain's Astronomer Royal on how the Universe is and why it is the way it is.

Percival Lowell (1895) *Mars*. Riverside Press, Cambridge MA.

A *fin de siècle* classic. The original and still the most entertaining work of speculative science concerning possible extraterrestrial life. It is now in the public domain and available online at http://www.wanderer.org/references/lowell/Mars/.

on cosmology, physics, parallel universes and the true nature of reality

Brian Greene (2000) *The Elegant Universe – Superstrings, Hidden Dimensions, and the Quest for the Ultimate Theory*. Random House, New York.

The first book to tackle the fiendishly complex world of superstring theory for a lay audience. Utterly brilliant.

Brian Greene (2004) *The Fabric of the Cosmos: Space, Time and the Texture of Reality*. Allen Lane, London.

Another gem from the man who has done more than any other, perhaps, to boil down the essence of the new physics and all its implications.

Stephen Hawking (1988) *A Brief History of Time: From the Big Bang to Black Holes*. Bantam Press, New York.

Everyone's got one, but how many copies are sitting on the bookshelves unread? A true classic that it is well worth persevering with – this book is not nearly as difficult a read as legend has it.

Sir Martin Rees (1999) *Just Six Numbers. The Deep Forces that Shape the Universe.* Weidenfeld & Nicholson, London.

Why is the Universe set up as it is? Why are the physical constants so finely tuned as to allow life to exist, at least in our tiny part of the cosmos? Rees tackles these questions and more in his entertaining and readable way.

Jim Al-Khalili (2003) *Quantum – A Guide for the Perplexed.* Weidenfeld & Nicholson, London.

Schrödinger's cat, 'spooky' action at a distance and quantum tele-portation: they are all here, in one of the best and most accessible accounts to date of a difficult, counter-intuitive science. As Niels Bohr said, 'Anyone who is not shocked by quantum theory has not under-stood it'. Read this and be shocked.

Michio Kaku (2005) *Parallel Worlds, The Science of Alternative Universes and our Future in the Cosmos.* Allen Lane, London.

A wholly accessible romp through the New Cosmology, dealing with everything you want to know about superstrings, parallel realities and possible escape routes for life in the far-distant future.

David Harland (2003) *The Big Bang: A View from the 21st Century.* Springer-Verlag, New York.

Britain's foremost space historian on the latest theories concerning the beginning of time.

on time machines

Paul Davies (2001) *How to Build a Time Machine.* Allen Lane, London.

Does what it says on the cover, provides a series of possible and plau-sible blueprints for a machine that could take you back to meet your grandfather – or indeed take you back in time to become him. Bril-liant writing by the British polymath working in Australia.

Julian Barbour (1999) *The End of Time. The Next Revolution in our Understanding of the Universe.* Weidenfeld & Nicholson, London.

A deeply strange, provocative and controversial thesis that states that time, as we understand it, is an illusion. If Barbour is right, just about everything we thought we knew about anything is wrong.

on the fate of the universe and its beginnings

Fred Adams and Greg Laughlin (1999) *The Five Ages of the Universe: Inside the Physics of Eternity.* Simon & Schuster, New York.

Highly entertaining account of the almost impossibly distant future, from the fate of the Earth five billion years hence (it fries) to the eventual death of the stars and even of matter itself. Adams and Laughlin also ponder the possibilities open to life as the Universe around it crumbles into dust.

on computing

John Naughton (1999) *A Brief History of the Future – The Origins of the Internet.* Weidenfeld & Nicholson, London.

Now slightly dated (this was written in the pre-Google age) this nevertheless remains the definitive account of the creation of the global computer network which has changed our civilization.

on god

Stephen Unwin (2004) *The Probability of God: A Simple Calculation That Proves the Ultimate Truth.* Three Rivers Press, New York.

Provocative reading for believers and unbelievers alike, Unwin comes up with the surprising conclusion that as a result of probability theory God probably exists.

Index